Lecture Notes in Computer Science 5907

Commenced Publication in 1973
Founding and Former Series Editors:
Gerhard Goos, Juris Hartmanis, and Jan van Leeuwen

Ryszard Kowalczyk   Quoc Bao Vo
Zakaria Maamar   Michael Huhns (Eds.)

Service-Oriented Computing: Agents, Semantics, and Engineering

AAMAS 2009 International Workshop, SOCASE 2009
Budapest, Hungary, May 11, 2009
Proceedings

 Springer

Volume Editors

Ryszard Kowalczyk
Swinburne University of Technology, John St, Hawthorn, VIC 3122, Australia
E-mail: rkowalczyk@ict.swin.edu.au

Quoc Bao Vo
Swinburne University of Technology, John St, Hawthorn, VIC 3122, Australia
E-mail: bvo@ict.swin.edu.au

Zakaria Maamar
CIT, Zayed University, Dubai, United Arab Emirates,
E-mail: zakaria.maamar@zu.ac.ae

Michael Huhns
University of South Carolina, Columbia, SC 29208, USA
E-mail: huhns@engr.sc.edu

Library of Congress Control Number: 2009939831

CR Subject Classification (1998): H.3.5, H.3.3, H.3-4, I.2, C.2.4

LNCS Sublibrary: SL 3 – Information Systems and Application, incl. Internet/Web and HCI

ISSN   0302-9743
ISBN-10   3-642-10738-9 Springer Berlin Heidelberg New York
ISBN-13   978-3-642-10738-2 Springer Berlin Heidelberg New York

springer.com

© Springer-Verlag Berlin Heidelberg 2009
Printed in Germany

Typesetting: Camera-ready by author, data conversion by Scientific Publishing Services, Chennai, India
Printed on acid-free paper   SPIN: 12800537   06/3180   5 4 3 2 1 0

Preface

The areas of service-oriented computing and semantic technology offer much of interest to the multiagent system community, including similarities in system architectures and provisioning processes, and powerful tools and standardizations to enable more flexible and dynamic business process integration and automation. Similarly, techniques developed in the multiagent systems and semantic technology areas are having a strong impact on the fast-growing service-oriented computing area. Other issues, such as quality of service, security, privacy, and reliability are common problems to both multiagent systems and service-oriented computing.

Service-oriented computing has emerged as an established paradigm for distributed computing and e-business processing. It utilizes services as fundamental building blocks to enable the development of agile networks of collaborating business applications distributed within and across organizational boundaries. Services are self-contained, platform-independent software components that can be described, published, discovered, orchestrated, and deployed for the purpose of developing distributed applications across large heterogeneous networks such as the Internet.

Multiagent systems, on the other hand, also aim at the development of distributed applications, however, from a different but complementary perspective. Service-oriented paradigms are mainly focused on syntactical and declarative definitions of software components, their interfaces, communication channels, and capabilities with the aim of creating interoperable and reliable infrastructures. In contrast, multiagent systems are focused on the development of reasoning and planning capabilities of autonomous problem solvers that actively apply behavioral concepts such as interaction, collaboration, and negotiation in order to create flexible and fault-tolerant distributed systems for dynamic and uncertain environments.

Semantic technology offers a semantic foundation for interactions among agents and services, forming the basis upon which machine-understandable service descriptions can be obtained, and as a result, autonomic coordination among agents is made possible. On the other hand, ontology-related technologies, ontology matching, learning, and automatic generation, etc., not only gain in potential power when used by agents, but also are meaningful only when adopted in real applications in areas such as service-oriented computing.

This volume consists of the proceedings of the Service-Oriented Computing: Agents, Semantics, and Engineering (SOCASE 2009) workshop held at the International Joint Conferences on Autonomous Agents and Multiagent Systems (AAMAS 2009). The papers in this volume cover a range of topics at the intersection of service-oriented computing, semantic technology, and intelligent multiagent systems, such as: service description and discovery; planning, composition and negotiation; semantic processes and service agents; and applications.

The workshop organizers would like to thank all members of the Program Committee for their excellent work, effort, and support in ensuring the high-quality program and successful outcome of the SOCASE 2009 workshop. We would also like to thank Springer for their cooperation and help in putting this volume together.

September 2009

Ryszard Kowalczyk
Quoc Bao Vo
Zakaria Maamar
Michael Huhns

Organization

SOCASE 2009 was held in conjunction with the 7th International Joint Conference on Autonomous Agents and Multiagent Systems (AAMAS 2009) on May 11, 2009 in Budapest, Hungary.

Organizing Committee

Ryszard Kowalczyk	Swinburne University of Technology, Australia
Quoc Bao Vo	Swinburne University of Technology, Australia
Zakaria Maamar	Zayed University Dubai, United Arab Emirates
Michael Huhns	University of South Carolina, USA

Program Committee

Jamal Bentahar	Concordia University, Canada
M. Brian Blake	Georgetown University, USA
Athman Bouguettaya	CSIRO, Australia
Jakub Brzostowski	Silesian University of Technology, Poland
Paul Buhler	College of Charleston, USA
Mauro Gaspari	Università di Bologna, Italy
Christian Guttmann	Monash University, Australia
Slimane Hammoudi	ESEO, France
Jingshan Huang	Benedict College, USA
Clement Jonquet	Stanford University, USA
Ryszard Kowalczyk	Swinburne University of Technology, Australia
Luis Llana	Universidad Complutense de Madrid, Spain
Zakaria Maamar	Zayed University Dubai, United Arab Emirates
Xuan Thang Nguyen	TIBRA, Australia
Manuel Nunez	Universidad Complutense de Madrid, Spain
Julian Padget	University of Bath, UK
Huaglory Tianfield	Glasgow Caledonian University, UK
Rainer Unland	University of Duisburg-Essen, Germany
Kunal Verma	Accenture, USA
Quoc Bao Vo	Swinburne University of Technology, Australia
Leandro Krug Wives	Federal University of Rio Grande do Sul, Brazil

Table of Contents

Service-Oriented Computing: Agents, Semantics, and Engineering

Contract Observation in Web Services Environments

Jiří Bíba, Jiří Hodík, Michal Jakob, and Michal Pěchouček

Agent Technology Center,
Dept. of Cybernetics, FEE, Czech Technical University,
Technická 2, 16627 Prague 6, Czech Republic
{hodik,jakob,pechoucek}@agents.felk.cvut.cz
http://agents.felk.cvut.cz

Abstract. Electronic contracting, based on explicit representation of different parties' commitments, is a promising way to specifying and regulating behaviour in distributed business applications. A key part of contract-based system is a process through which the actual behaviour of individual parties is checked for conformance with contracts set to govern such behaviour. Such checking requires that relevant information on the behaviour of the parties, both with respect to the application processes they execute and to managing their contractual relationships, is captured. The process of collecting all such information, termed *contract observation*, is the subject of this paper. First, we describe general properties and requirements of such an observation process; afterwards, we discuss specifics of realising contract observation in web services environments. Finally, we show how contract observation has been implemented as part of the IST-CONTRACT web services framework for contract-based systems.

1 Introduction

Of the ways in which agent behaviour can be regulated in a multi-agent system, electronic contracting – based on explicit representation of different parties' commitments and the agreement of all parties to them – has significant potential for modern distributed applications. In part, this is because it explicates different parties' responsibilities, and the agreement of all parties to them, allowing businesses to operate with expectations of the behaviour of others, but providing flexibility in how they fulfil their own obligations. Additionally, it mirrors existing (non-electronic) practise, aiding adoption.

A key element of contract-based systems is *contract monitoring*, i.e. a process by which the behaviour of individual parties and their participation in respective business processes is checked for compliance with contracts set to regulate such behaviour. The term monitoring has been traditionally used to refer both to the process through which contract-relevant events and states from a running contract-based distributed system are gathered, and the reasoning applied over the gathered information in order to determine the compliance with the

R. Kowalczyk et al. (Eds.): SOCASE 2009, LNCS 5907, pp. 1–11, 2009.
© Springer-Verlag Berlin Heidelberg 2009

respective contracts. Although such tight coupling can have certain advantages, it limits the ways in which both processes can be configured and combined to meet distinct requirements of different application domains.

In our approach, we therefore separate the data gathering process, termed *contract observation* further on, and study it independently of other operations and processes required for contract monitoring.

Specifically, in Section 2 we analyse the observation process in general, primarily from the perspective of the categories of information that need to be observed in order to determine fulfilment state of contracts. In Section 3, we continue by exploring different ways in which the general observation process can be realised in a web services environment. In Section 4 we further narrow down our scope and describe how contract observation has been implemented in the IST-CONTRACT middleware [1]. In Section 5 we discuss related work and we conclude in Section 6.

2   Contract Observation Process

In general, a *contract* is a set of restrictions on the behaviour of involved contract parties. Such restrictions may involve both achieve and maintain conditions, e.g. a prohibition to perform a particular operation or an obligation to keep a certain quantitative measure of agent's operation within prescribed limits, respectively. The behaviour of contract parties can be modelled as composed of elementary units of activity, termed *actions* further on. Contract behaviour restrictions can then be defined in terms of a set of contract clauses specifying which action is required, prohibited or allowed for contract parties under which conditions. These conditions usually refer to the state of the environment within which the parties operate (termed *contracting environment*) but can also refer to the state of other contracts. See [2] for a detailed discussion of contract structure and semantics.

In order to be able to determine fulfilment of a contract, it is necessary to observe information at two levels:

- information about the actions carried out by parties involved in the contract and the state of the contracting environment (*domain-level information*)
- information about the content and life-cycle status of the contract (*contract-level information*)

Whereas domain-level information is required to track agent behaviour (which is subject to contract regulation), contract-level information is necessary to track which contracts are currently active and against which the behaviour of contract parties should thus be checked for compliance.

Observing domain-level actions of a contract party agent requires gathering information on how the agent manipulates the environment in which the contract-regulated process executes. In the case of electronic systems, such an environment comprises a universe of electronic resources; observing agent's actions then correspond to the logging of operations on those resources (e.g. delivering a required file, transferring an agreed sum, launching a specified process

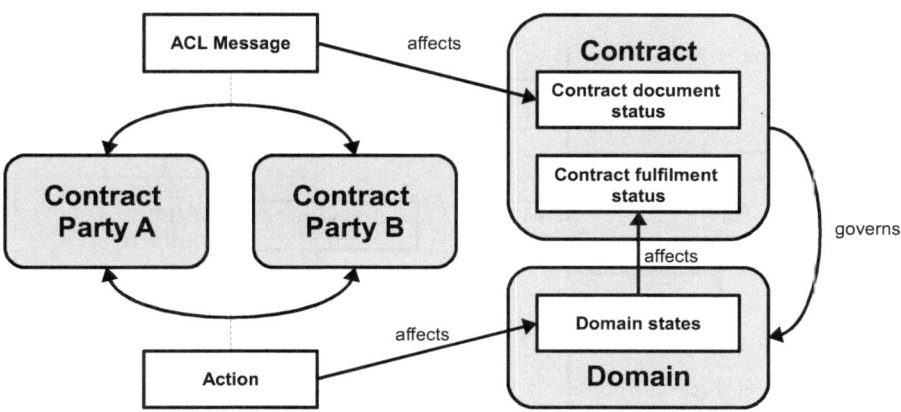

Fig. 1. Observation in contract-based systems. Actions are observed at the domain level; contract-affecting communication (ACL messages) is observed at the contract level. The information is processed by the monitor to determine the contract lifecycle and fulfillment status.

etc.). In order to allow to determine the compliance of a particular observed action, it is necessary that the observation process provides, at minimum, (i) the identity of the agent performing the action, and (ii) detailed parameters of the action (e.g. the sum and the destination bank account in the case of the money transfer action).

In contrast, changes to the content (adding, removing or modifying contract clauses) and life-status of contracts (signature, termination, renewal etc) take place through direct interaction of the respective contract parties. Observing contract-level information thus requires monitoring the communication between the agents and logging any contract-affecting operations. The relation between the processes in a contract-based system and the different categories of information observed is depicted graphically in Figure 1.

Once the required information from the system is obtained, the actual decision on the contract compliance has to be made. This is generally a non-trivial operation requiring a reasoner capable of interpreting deontic concepts. Due to our focus on the observation, a module capable of such operation, termed *monitor*, is viewed as a blackbox and its implementation is not discussed in detail in this paper. We assume that the input to the monitor contains both domain- and contract-level information, specifically the state of the contracting environment, actions performed by the contract parties, contracts between the parties and the life-cycle status of the contracts (in particular whether any given contract is active). In its basic form, the monitor outputs, for each contract and/or each contract clause, the information whether the contract and/or the clause is being violated. We assume that the monitor operates sequentially and is causal in a sense that for determining the fulfilment state in a particular point in time it only needs information observed up to that point in time.

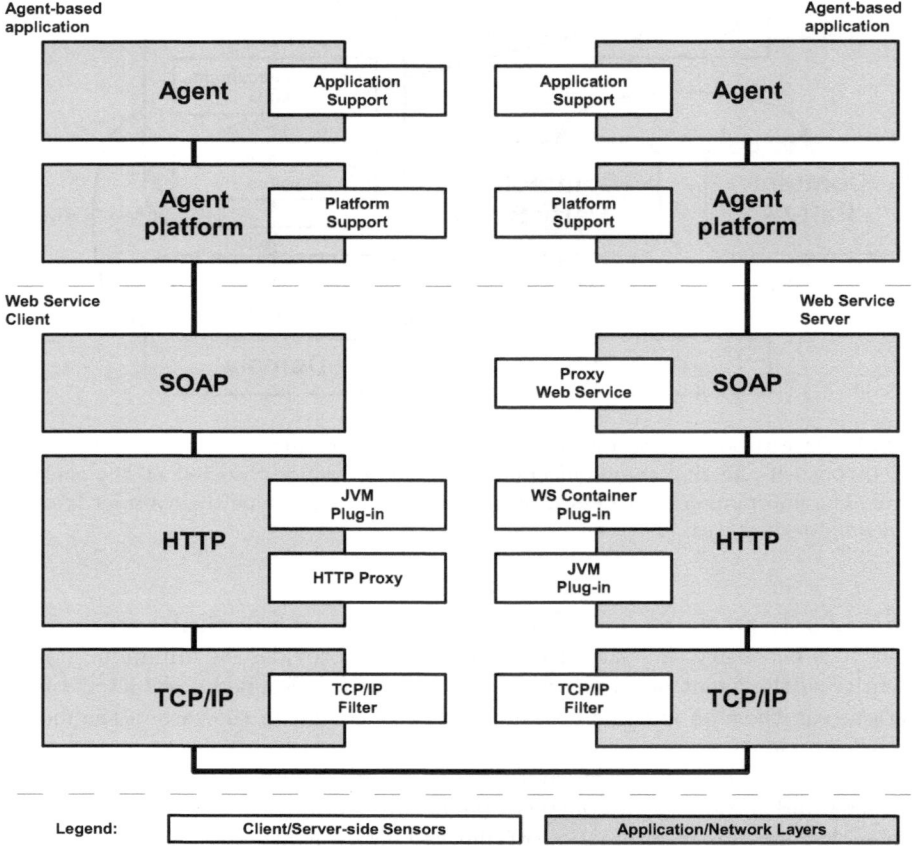

Fig. 2. Possible implementations of observation sensors in the WS-based contract party interaction stack

3   Observation in Web Services Environments

We now show how the abstract observation processes described in the previous section can be realized in web services (WS) environments. The basic assumption made is that execution of an action in WS environments corresponds to the invocation of a particular web service.

In general, any interaction between a contract party agent and other agents and/or resources deployed in a WS environment passes through several layers. It starts as an invocation of the agent platform API, continuing down the architecture through the agent platform messaging layer using a WS stack which, translating the invocation by a SOAP processor, creates an HTTP message sent to the other interacting party and/or resource using a TCP/IP connection – see Figure 2.

In theory, interactions can be intercepted on any of the above levels. In the following, we assume that the interception is performed by a special software

Table 1. Comparison of different sensor implementations

	Advantages (+)	Disadvantages (−)
application-level API	− all information needed for monitoring is instantly available	− reporting of selected actions and/or communication can be (intentionally) omitted
agent platform plug-in	− all information needed for monitoring is instantly available − reporting cannot be circumvented	− communication and action execution reporting has to be supported by the agent platform
proxy web service	− the proxy web service code can be extended by information needed for monitoring while the target web service remains intact and operational − platform/OS independent	− lack of stable open-source tools and frameworks implementing this functionality − may be a performance bottleneck
JAX-WS/JVM plug-in	− transparent for both client and server solutions (possibly using the same code) − does not require any modifications to existing services and/or clients	− the extension/modification may not be feasible − deployment on web service/servlet containers may be difficult
web service/servlet container plug-in	− usable transparently at the server side, uses container API − does not require any modifications to existing services and/or clients	− container dependence (not all containers implement the necessary JSR specifications)
http proxy	− usable transparently on the client side − minimal changes to existing web service clients (only configuration) − platform/OS independent	− more http proxies for more JVMs on the same computer may pose performance problems (stability, robustness, load)
TCP/IP filter	− transparent for both client and server solutions − does not require any modifications to existing services and/or clients	− platform/OS dependent − may rise security issues and conflicts with existing software operating at the TCP/IP level

module termed *sensor* which extracts the relevant data and sends them to a designated collection point. Specifically, we consider the following sensor insertion points (see Figure 2 for a graphical overview):

application-level API [*both client and server side*]: the sensor is integrated into the contract party application code and explicitly called to report inter-agent messaging and/or action invocation

agent platform plug-in [*both client and server side*]: the sensor is integrated
  into the agent platform so that all communication and action invocation is
  automatically reported

proxy web service [*server side only*]: the target web service is encapsulated
  within a proxy web service mirroring the target web service operations; the
  proxy is invoked instead of the target web service and the sensor is integrated
  into the code of the proxy service (the proxy web service can be generated
  based on the WSDL of the target web service)

JAX-WS/JVM plug-in [*both client and server side*]: the SOAP processor in
  the JAX-WS library is extended to integrate the sensor or the sensor is
  plugged directly into the JVM; in either case, the container hosting the web
  services needs to use such a modified JAX-WS library or JVM instead of
  the standard distributions

web service/servlet container plug-in [*server side only*]: the target web service deployed in a web service container is observed using mechanisms offered
  by (some) servlet containers (e.g. Glassfish implements the necessary JSR
  monitoring and management specifications to intercept HTTP traffic between the container (server) and its clients) and a public API to attach the
  sensor to is available

http proxy [*client side only*]: all outbound traffic is filtered at the HTTP layer
  by a proxy application which filters SOAP calls and synthesise the required
  information

TCP/IP filter [*both client and server side*]: all traffic is filtered at the TCP/IP
  layer by means of a special application (usually integrated directly into the core
  of the underlying operating system) and HTTP messages containing SOAP
  calls are identified and used to synthesise the information to be reported

Table 1 summarises the advantages and disadvantages of different sensor implementations.

4   Observation in the CONTRACT Framework

In this section, we describe how contract observation has been implemented in
the CONTRACT Framework developed within the IST-CONTRACT project[1].
We first overview the architecture of the whole framework and then describe the
implemented observation process.

4.1   CONTRACT Framework Architecture

CONTRACT is a WS framework for developing, implementing and monitoring
contract-based systems. The core of the framework is a JAX-WS compliant
web-service-based agent platform[2], also developed within the IST-CONTRACT
project. Agents are implemented as stateful web services which are accessed

---

[1] http://ist-contract.org
[2] Available from http://ist-contract.sourceforge.net/

by means of a single stateless factory entry point returning a WS-Addressing compliant reference used for the invocation of individual agent operations. The web-service interface of each agent offers means for FIPA-ACL [3] compliant interactions between agents as well as operations for connecting the agent with external components such as a graphical front-end etc.

4.2   Observation Pipeline

The observation process in the CONTRACT framework is referred to as the *observation pipeline* and is realized through a distributed collaboration of several different types of agents. After initial research and experimentation, the agent platform plug-in option has been chosen for implementing the observation gathering sensors (see Section 3 for details). The decision was made because of continuing difficulty to find a solution at a different level that would work reliably across a wide range of deployment scenarios supported. Agent communication and action selection modules have been therefore equipped with a sensor that reports any communication and action invocation performed by the agent.

Altogether, the following components are involved in the CONTRACT observation pipeline:

– **Sensor** – an interface implemented by the Contract Party to allow observation of its contract-related activities. The Sensor provides domain- and contract-level data to the Observer agent. The main functionality of Sensors,

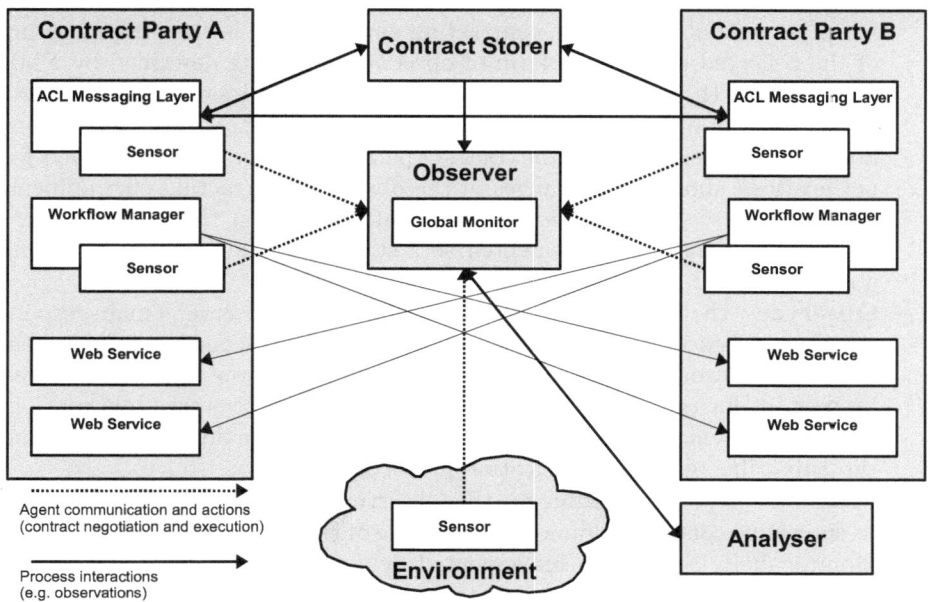

Fig. 3. Architecture of the CONTRACT framework

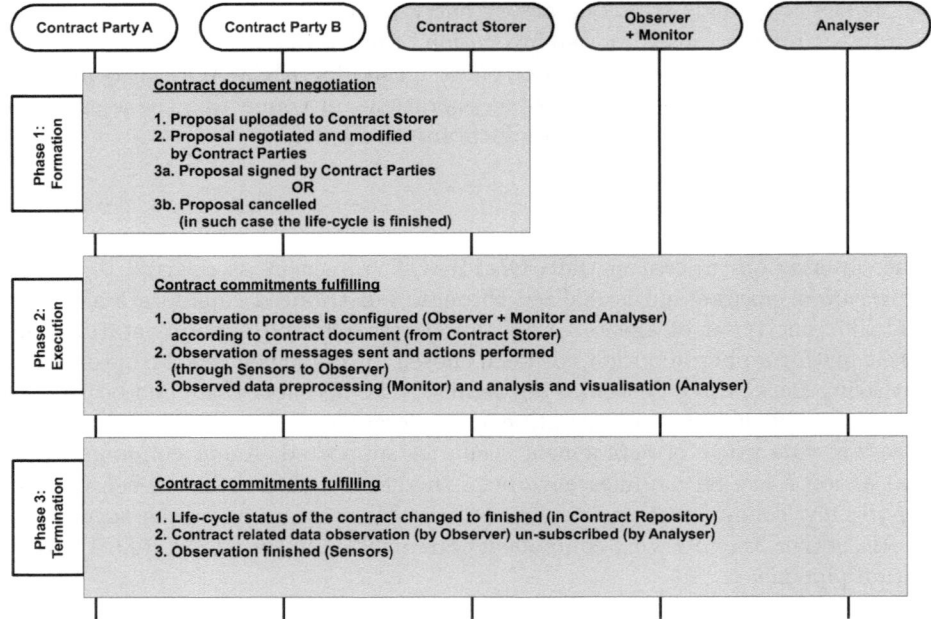

Fig. 4. Interactions between the CONTRACT framework components during individual stages of the contract life-cycle

which are distributed in the contracting environment, is the pre-processing of the collected data into a form of observation reports described by XML schemas and the submission of these reports to the designated Observers. Three types of observation reports are used: *action reports* for notifications about actions performed by the contract parties, *domain predicate reports* for notifications about state changes of the observed contracting environment, and *ACL message reports* for communication between the contract party agents. The Sensor is implemented as a standalone Java library providing a reporting API.

- **Observer** – the central component of the pipeline. Observer's main responsibility consists in collecting data reported by sensors and providing them further to the other pipeline components. The Observer may by extended by plug-ins for the pre-processing and processing of the stored information. Observer provides an ACL-compliant interface supporting the Query and the Subscribe agent communication protocols.
- **Monitor** – a plug-in module for the Observer processing the observed data to determine contract fulfilment. In the case of the CONTRACT project, the Monitor module has been implemented by a means of a reasoner based on augmented transition networks (see [4] for details). As its input, the Monitor receives the observation reports gathered by the encompassing Observer; the output of the Monitor (the fulfilment state of all monitored contracts and

their clauses) is forwarded to the Observer from which it is made available to other components in the system.
– **Contract Storer** – the agent keeping track of contracts in the system, including their life-cycle status (in cooperation with the Observer). The contract storer acts as an authoritative source of information on contract content and life-cycle status, which is used by other components of the pipeline. Internally, the Contract Storer agent uses an eXist XML database to store contract documents.

In addition, there is the Analyser agent providing a user front-end to the pipeline, presenting the contract status and fulfilment data to the administrators of the monitored contract-based system. Interactions between the components within the CONTRACT Framework are depicted in Figure 3; their involvement in individual stages of the contract life-cycle is depicted in the Figure 4.

The concept of presented pipeline and the underlying WS framework has been validated on several use-cases, including modular certification testing, insurance claim handling and the maintenance and support of aircraft engine units (see [5] for the description of the use cases).

5   Related Work

This work focuses on the observation and monitoring of distributed systems whose operation can be modelled as a choreography of well-defined elementary actions. This contrasts with most of the work on monitoring service level agreements (e.g. [6,7,8]) which focus on continuous evaluation of a set of performance metrics and their comparison with agreed thresholds.

Approaches more relevant to our work range from theoretical concepts establishing service frameworks for contract descriptions and monitoring (e.g. [9,10]) to implementations of tools and other means for run-time monitoring of workflow executions based on electronic contracts as declarative specifications of multi-party cooperative behaviours (such as [11,12,13,14]).

Mahbub and Spanoudakis [11] use event calculus for defining monitoring requirements on top of a workflow described in BPEL4WS [15]. The behaviour requirements are automatically extracted from the workflow description and can be extended in order to describe an overall behaviour of a service composition in terms of temporal constraints and properties of the data processed during web service invocations.

Barbon et Al. [13] present a monitoring module extending the open-source Active BPEL workflow engine. The monitoring module intercepts events in the workflow life-cycle (creation and termination) and invocations of external services; based on the recorded data, the monitor checks for run-time errors, timeouts, and functional properties. The monitoring specification is expressed as logical formulas in the RunTime Monitor specification Language (RTML) and can be automatically translated into a Java code implementing the monitor functionality. The monitoring logic is kept separate from the BPEL process (no modification required) but cannot be deployed in other BPEL engines.

Another approach for monitoring workflows is presented by Baresi et Al. [12]. They provide a tool for annotating workflows with assertions (monitoring rules) described in a proprietary Web Service Constraint Language (WS-CoL) being inspired by the JML [16]. Their approach clearly separates the original business logic from the superimposed monitoring code and is independent of any specific workflow engine.

All of the above work, however, views observation as an inseparable part of the complete monitoring solution. Modular approach taken in this paper, viewing observation as an abstract process with multiple possible implementations, is to our best knowledge novel.

6   Conclusions

Observation, i.e. the process of obtaining the relevant information on the operation of a distributed business system, is a key pre-requisite for determining whether the system operation complies with contracts set to regulate it. The observation has to ensure that sufficient information on the action of contract parties in the system is recorded and provided in a suitable form. In this paper, we identified the type and form of information required and analysed possible ways in which the information can be obtained in web-services based environments. We then showed a particular implementation of the observation process, as implemented by the IST-CONTRACT project in a web services-based framework for contract-based systems.

Acknowledgements

The research described is part-funded by the EC FP6 projects CONTRACT (contract No. 034418), I*PROMS Network of Excellence, and also by the Ministry of Education, Youth and Sports of the Czech Republic grant No. MSM 6840770038. The opinions expressed herein are those of the named authors only and should not be taken as necessarily representative of the opinion of the European Commission or CONTRACT project partners.

References

1. Confalonieri, R., lvarez Napagao, S., Panagiotidi, S., Vzquez-Salceda, J., Willmott, S.: A middleware architecture for building contract-aware agent-based services. In: Kowalczyk, R., Huhns, M.N., Klusch, M., Maamar, Z., Vo, Q.B. (eds.) Service-Oriented Computing: Agents, Semantics, and Engineering. LNCS, vol. 5006, pp. 1–14. Springer, Heidelberg (2008)
2. Oren, N., Panagiotidi, S., Vazquez-Salceda, J., Modgil, S., Luck, M., Miles, S.: Towards a formalisation of electronic contracting environments. In: COIN 2008: Proceedings of AAAI Workhop on Coordination, Organization, Institutions and Norms in Agent Systems (2008)

3. FIPA: Foundation for intelligent physical agents (December 2003), http://www.fipa.org
4. Faci, N., Modgil, S., Oren, N., Meneguzzi, F., Miles, S., Luck, M.: Towards a monitoring framework for agent-based contract systems. In: Klusch, M., Pĕchouček, M., Polleres, A. (eds.) CIA 2008. LNCS (LNAI), vol. 5180, pp. 292–305. Springer, Heidelberg (2008)
5. Jakob, M., Miles, S., Luck, M., Oren, N., Kollingbaum, M., Holt, C., Vazquez, J., Storms, P., Dehn, M.: Case Studies for Contract-based Systems. In: Proceedings of the 7th International Conference on Autonomous Agents and Multiagent Systems (2008)
6. Keller, A., Ludwig, H.: The WSLA Framework: Specifying and Monitoring Service Level Agreements for Web Services. Journal of Network and Systems Management 11(1), 57–81 (2003)
7. Andrieux, A., Czajkowski, K., Dan, A., Keahey, K., Ludwig, H., Pruyne, J., Rofrano, J., Tuecke, S., Xu, M.: Web Services Agreement Specification (WS-Agreement). In: Global Grid Forum GRAAP-WG, Draft (August 2004)
8. Schopf, J., Raicu, I.: Pearlman Let al. Monitoring and discovery in a web services framework: Functionality and performance of Globus Toolkit MDS4. Technical report, Technical Report, Mathematics and Computer Science Division, Argonne National Laboratory (2006)
9. Xu, L., Jeusfeld, M.: Pro-active Monitoring of Electronic Contracts. In: Advanced Information Systems Engineering, pp. 584–600. Springer, Heidelberg (2003)
10. Milosevic, Z., Gibson, S., Linington, P., Cole, J., Kulkarni, S.: On design and implementation of a contract monitoring facility. In: Proceedings of the First IEEE International Workshop on Electronic Contracting, pp. 62–70 (2004)
11. Mahbub, K., Spanoudakis, G.: A framework for requirents monitoring of service based systems. In: Proceedings of the 2nd international conference on Service oriented computing, pp. 84–93. ACM New York, NY (2004)
12. Baresi, L., Guinea, S.: Towards Dynamic Monitoring of WS-BPEL Processes. In: Benatallah, B., Casati, F., Traverso, P. (eds.) ICSOC 2005. LNCS, vol. 3826, pp. 269–282. Springer, Heidelberg (2005)
13. Barbon, F., Traverso, P., Pistore, M., Trainotti, M.: Run-time monitoring of instances and classes of web service compositions. In: ICWS, vol. 6, pp. 63–71
14. Radha Krishna, P., Karlapalem, K., Chiu, D.: An EREC framework for e-contract modeling, enactment and monitoring. Data & Knowledge Engineering 51(1), 31–58 (2004)
15. OASIS: Oasis web services business process execution language (wsbpel) Web Services Business Process Execution Language Version 2.0. Committee Draft (May 2006), http://www.oasis-open.org/committees/download.php/18714/wsbpelspecification-draft-May17.htm
16. Leavens, G., Baker, A., Ruby, C.: Preliminary design of JML: a behaviora⌐ interface specification language for java. ACM SIGSOFT Software Engineering Notes 31(3), 1–38 (2006)

Mechanism Design for Task Procurement with Flexible Quality of Service

Enrico H. Gerding[1], Kate Larson[2], Alex Rogers[1], and Nicholas R. Jennings[1]

[1] University of Southampton, Southampton, SO17 1BJ, UK
{eg,acr,nrj}@ecs.soton.ac.uk
[2] University of Waterloo, Waterloo, ON N2L 3G1, Canada
klarson@cs.uwaterloo.ca

Abstract. In this paper, we consider the problem where an agent wishes to complete a single computational task, but lacks the required resources. Instead, it must contract self-interested service providers, who are able to flexibly manipulate the quality of service they deliver, in order to maximise their own utility. We extend an existing model to allow for multiple such service providers to be contracted for the same task, and derive optimal task procurement mechanisms in the setting where the agent has full knowledge of the cost functions of these service providers (considering both simultaneous and sequential procurement). We then extend these results to the incomplete information setting where the agent must elicit cost information from the service providers, and we characterise a family of incentive-compatible and individually-rational mechanisms. We show empirically that sequential procurement always generates greater utility for the agent compared to simultaneous procurement, and that over a range of settings, contracting multiple providers is preferable to contracting just one.

1 Introduction

Service-oriented computing, in which computational resources are seamlessly and dynamically procured from third party suppliers as they are required, has generated significant recent activity within the research community. Examples of such initiatives include Grid and utility computing, and these technologies are increasingly being proposed for a wide range of scientific and business workflows. However, to reach the full potential of this vision, such systems require that both the suppliers and consumers of these computational resources are able to engage in autonomous negotiation and contracting (given their own individual goals and requirements). To this end, agent-based approaches that make use of computational mechanism design have been advocated [1].

Much of the work in this area to date has sought to extend the standard approaches of mechanism design to cases in which there is a non-zero probability that a service provider may fail to meet its contracted obligation (e.g. failing to satisfy an agreed time deadline) [2,3]. However, a significant shortcoming of much of this work is that it assumes the probability with which a service provider will fail is fixed and exogenous. In contrast, Matsubara provides a more realistic model in which self-interested service providers flexibly manipulate their quality of service in order to maximise their own utility [4]. In this model, a contracted service provider actively manages the resources

R. Kowalczyk et al. (Eds.): SOCASE 2009, LNCS 5907, pp. 12–23, 2009.

that it commits to a task, in order to manipulate the quality of service that it provides to the contracting agent (e.g., it may commit more resources to a task, increasing the probability that the task will be successfully completed, if the reward for doing so is large; conversely, it may intentionally fail to complete a task if it profits by doing so). Here the contracting agent is faced with a principal-agent problem since it cannot directly monitor the service providers' activities. Hence it must create appropriate incentives in order to achieve the desired quality of service level.

To this end, Matsubara provides a mechanism based on a payment rule that incentivises potential contractors to truthfully reveal their costs for completing the task, and, once selected, to invest the required amount of costly resources. The approach used is similar to that more recently applied to an information setting where *strictly proper scoring rules* are used to incentivise information providers to truthfully reveal a probabilistic estimate whose generation requires the investment of costly resources [5,6]. However, these mechanisms are restricted to the case that a single service provider is contracted to complete each task. In reality, when faced with the uncertain execution of tasks, it is common to introduce redundancy by contracting multiple providers for the same task; either simultaneously or sequentially (i.e. the agent awaits the failure of one provider before approaching another) [7].

It is this shortcoming that we address in this paper, and to this end, we describe a novel family of mechanisms that allow the contracting agent to procure computational tasks from multiple self-interested service providers. We consider how a utility-maximising contracting agent should price contracts with these service providers in both the full information setting, where the agent has complete knowledge of the cost functions of the providers, and also in the incomplete information setting where it must elicit this information from them. Such an extension is challenging since, in the optimal case, the expected utility of a provider that has been contacted to perform the task, depends on the cost functions of the other contracted providers. In the incomplete information case this results in an interdependent valuation setting with so-called allocative externalities, for which it has been shown that no standard mechanism exists which is both efficient and incentive compatible [8]. We address this by developing mechanisms which are not efficient, but take advantage of the optimal solution to reduce this inefficiency. We then show empirically that, under our mechanisms, contracting multiple service providers for the same task is preferable to contracting a single provider, as it often increases the probability that a task is successfully completed, whilst reducing the costs to the contracting agent.

In more detail, we make the following contributions to the state of the art:

- We derive optimal task procurement mechanisms when the agent has full knowledge of the cost functions of the service providers. We consider procurement from multiple providers, and consider settings where these providers are contracted simultaneously and sequentially (i.e. the agent awaits the failure of one provider before approaching another).

- We extend these results to the incomplete information setting where the agent must elicit the cost information from the providers. We characterise a family of mechanisms and prove that they are incentive compatible (i.e. the providers have a dominant strategy to truthfully reveal their cost functions to the agent), and individually rational (i.e. the expected utility of providers that participate in the mechanism is greater or equal to zero). Based on the insights obtained from the optimal, full information case, we present three mechanisms from this family: a uniform and discriminatory pricing mechanism for the simultaneous procurement case, and a mechanism for the sequential procurement case.

- Finally, we empirically evaluate our mechanisms, showing that sequential procurement always generates greater utility for the agent compared to simultaneous procurement, that discriminatory pricing always generates greater utility for the agent than uniform pricing, and that over a range of settings, procuring from multiple providers is preferable to procuring from just one.

The remainder of this paper is structured as follows. In section 2 we formally describe the setting which we consider. In section 3 we describe the optimal procurement strategy in the full information setting, before presenting our three novel mechanisms, and proving their properties, in section 4. We instantiate and empirically evaluate these mechanisms in section 5, and, conclude in section 6.

2   Problem Description

Our model is based on that of Matsubara, but expressed here in the standard notation of mechanism design, rather than the original contract theory. Hence, we consider that contracting agent, A, has a task, T, that it would like to have completed. If the task is completed successfully then the agent receives value V, and otherwise it receives zero. We assume that there are n service providers capable of performing task, T. The probability of any provider successfully completing the task (the quality of service offered) depends on the amount of some costly resources that it decides to allocate to the task.

Formally, we assume that each service provider, i, has a potentially unlimited supply of resources[1], and denote $r_i \geq 0$ as the amount of resources that i will devote to executing the task. We assume that as a provider allocates more resources to the task, the probability that the task will be successfully executed increases. That is, there is a *quality of service* (QoS) function $P : \mathbb{R}^+ \to [0, 1]$ such that $P(r_i)$ is the probability that i successfully completes the task if it devotes r_i resources to the problem. We assume that $P(\cdot)$ is common to all providers, that if a provider devotes *no* resources to the task then it will fail (that is, $P(0) = 0$), and that the more resources are devoted to the task, the more likely it is to successfully complete the task. Thus, $P(\cdot)$ is continuous, increasing and strictly concave with $P(r_i) \to 1$ as $r_i \to \infty$. Finally, we assume that the probability of success of any service provider depends only on its own resource allocation, and not on the success or failure of any other provider.

We model the costly resources of provider i with a cost function, $c_i : \mathbb{R}^+ \to \mathbb{R}$. We assume that $c_i(\cdot)$ is continuous, increasing and convex, and that $c_i(0) = 0$. In addition, we assume that for any two service providers i and j, if they have different cost functions, $c_i(r)$ and $c_j(r)$, these functions are non-crossing for $r > 0$, i.e., either $c_i(r) = c_j(r), \forall r$ or $c_i(r) \neq c_j(r), r > 0$. An example class of cost functions which

---

[1] However, the *costs* of these resources (explained below) can become arbitrarily large.

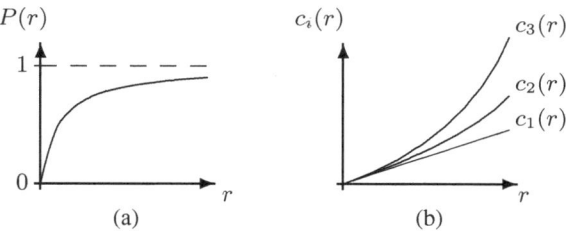

Fig. 1. Figure (a) is an example quality of service function. Figure (b) shows example cost functions.

satisfy these properties are *linear* cost functions, where $c_i(r_i) = K_i r_i$ for some constant $K_i \geq 0$. Figure 1 illustrates the structure of possible cost and quality of service functions.

Since the contracting agent cannot directly observe the amount of invested resources (but only whether a task failed or succeeded), it must create incentives for service providers to invest a certain amount of resources. To do so, the contracting agent uses a payment scheme whereby the payment depends on whether or not the task was successfully completed. In the case that service provider i is contracted, it is automatically paid β_i, and then, if the task is successfully completed, it receives an additional bonus α_i. We assume that α_i is always non-negative, but place no restrictions on β_i.

3   The Full Information Setting

In this section we study the problem of procuring service providers to complete the agent's task, when the agent has full information about the providers' cost functions. We first analyse the case where the agent contacts a single provider, before extending these results to the case where the agent procures services from multiple providers.

3.1   Single Service Provider Case

Since the providers are self-interested and autonomous, the agent is unable to force them to execute the task. Instead, the agent must provide appropriate incentives so that the providers will take on the task for the agent, and will invest appropriate levels of resources. Assume that the agent has selected provider i to execute the task. By setting the parameter α_i appropriately, the agent is able to induce any desired level of effort, $\bar{r}_i$, from provider i.[2] In particular, for any value of $\bar{r}_i$, provider i's expected utility is given by $U_i(\bar{r}_i) = \alpha_i P(\bar{r}_i) + \beta_i - c_i(\bar{r}_i)$ which is maximized when $U_i'(\bar{r}_i) = \alpha_i P'(\bar{r}_i) - c_i'(\bar{r}_i) = 0$. Solving for α_i, we have:

$$\alpha_i = \frac{c_i'(\bar{r}_i)}{P'(\bar{r}_i)} \tag{1}$$

---

[2] From here on, we use $\bar{r}_i$ to denote the *agent*'s desired level of effort, to distinguish from the *provider*'s chosen investment r_i, and to emphasize that the agent cannot directly enforce this.

which is well defined (since $c_i(\cdot)$ is convex and $P(\cdot)$ is strictly concave), and is positive (since both $c_i(\cdot)$ and $P(\cdot)$ are increasing). Therefore, if the agent desires that i invest $\bar{r}$ resources, then by setting $\alpha_i = c_i'(\bar{r})/P'(\bar{r})$, i maximizes its utility by actually investing $\bar{r}$ (i.e. $\bar{r} = argmax_r U_i(r)$).

As well as inducing effort, the agent would like to minimise the payment to provider i. At the same time, however, the agent needs to ensure that i will voluntarily participate, and as such must ensure that $U_i(\bar{r}_i) \geq 0$. The agent can satisfy both conditions by ensuring that $\beta_i \geq c_i(\bar{r}_i) - \alpha_i P(\bar{r}_i)$. By substituting α_i from Equation 1 and minimising β_i subject to the constraint, we have:

$$\beta_i = c_i(\bar{r}_i) - \frac{c_i'(\bar{r}_i)}{P_i'(\bar{r}_i)} P(\bar{r}_i). \tag{2}$$

While the agent wants to ensure that the selected provider invests the desired level of resources into the task, its real goal is to maximise its own utility, $U_A(\bar{r}) = (V - \alpha_i)P(\bar{r}) - \beta_i$. To this end, the agent must find α_i^* and β_i^* so as to induce the optimal level of effort from i, $\bar{r}_i^*$, such that $U_A(\bar{r}_i^*)$ is maximised. By substituting Equations 1 and 2 into the expression for the agent's utility, we get $U_A(\bar{r}_i) = VP(\bar{r}_i) - c_i(\bar{r}_i)$. Hence, by taking the first derivative and setting it to zero, and by letting $\bar{r}_i^*$ denote the optimal level of (induced) investment by provider i from the agent's perspective, we get $V = c_i'(\bar{r}_i^*)/P'(\bar{r}_i^*)$ and therefore:

$$\alpha_i^* = V = \frac{c_i'(\bar{r}_i^*)}{P'(\bar{r}_i^*)} \tag{3}$$

To calculate β_i^*, let $g(r_i) = c_i'(r_i)/P'(r_i)$. Then $\bar{r}_i^* = g^{-1}(V)$ and thus:

$$\beta_i^* = c_i(g^{-1}(V)) - VP(g^{-1}(V)). \tag{4}$$

It is important to note that when V (and thus α_i) is very small, and the provider's marginal costs $c'(r)$ are very high, the optimal level of effort $\bar{r}_i^* = g^{-1}(V)$ may be negative. If this is the case, i cannot hope to obtain any (strictly) positive utility, irrespective of the actual invested effort. However, due to voluntary participation, the provider will then choose to not execute the task, and this is actually beneficial to the agent since costly providers will self-select and voluntarily opt out.

3.2   Multiple Service Providers Case

By procuring from multiple service providers, the agent may be able to increase the probability with which the task is successfully executed since if one of them fails, then another may succeed. We now consider two procurement strategies involving multiple service providers: (i) simultaneous procurement and (ii) sequential procurement (where the agent awaits the failure of one provider before approaching another).

Simultaneous Procurement. In simultaneous procurement, the agent contracts some subset of providers, M with $|M| = m \leq n$, to execute the task T. Once contracted, the m providers all execute the task at the same time, not waiting to see how others perform. It may so happen that while the agent only requires the task to be completed once, multiple providers successfully complete the task. Given this possibility, the agent is faced with the problem of determining how to set the parameters α_i and β_i for each of the m providers in order to maximise its own utility.

Let $\bar{\mathbf{r}} = (\bar{r}_1, \ldots, \bar{r}_n)$ be the resource vector specifying (induced) resource allocations for each provider i. Assume that $r_i = 0$ for all $i \notin M$. The expected utility of the agent, given that it selected the m providers to procure from, is:

$$U_A^{\text{sim}}(\bar{\mathbf{r}}, M) = V \left(1 - \prod_{i \in M} [1 - P(\bar{r}_i)] \right) - \sum_{i \in M} [\alpha_i P(\bar{r}_i) + \beta_i]$$

$$= V \left(1 - \prod_{i \in M} [1 - P(\bar{r}_i)] \right) - \sum_{i \in M} c_i(\bar{r}_i). \tag{5}$$

Here, the latter equation is obtained by replacing α_i and β_i with Equations 1 and 2. By doing so, the agent induces investment $\bar{r}_i$, and in addition ensures voluntary participation. To determine the optimal investment levels for each provider i, $\bar{r}_i^*$, from the agent's perspective, we must solve $\nabla U_A^{\text{sim}}(\mathbf{r}, M) = 0$. This results in:

$$\alpha_i^* = V \prod_{j \neq i \in M} (1 - P(\bar{r}_j)). \tag{6}$$

The optimal solution $\bar{\mathbf{r}}^*$ and α_i^* for all i is found by solving the system of equations characterised by Equation 6, subject to the constraint that $\bar{r}_i^* \geq 0$ and $U_i(\bar{r}_i^*) = \alpha_i P(\bar{r}_i^*) + \beta_i^* - c_i(\bar{r}_i^*) \geq 0$. We note that, although at first glace it appears as though α_i^* does not depend on $c_i(\cdot)$, due to the interaction between the constraints for all providers, α_i^* is indirectly dependent on $c_i(\cdot)$. While this observation is interesting but irrelevant in the case where the cost functions are known, it will become important when we consider a mechanism for the incomplete information case in the next section.

Sequential Procurement. In sequential procurement, the agent only contracts an additional provider if previously contracted providers have failed to complete the task. This approach can be advantageous to the agent as it may only need to contract a small number of service providers, compared to the simultaneous procurement setting. On the other hand, the agent must decide the order in which to contract with providers.

The problem of sequentially procuring tasks from a set of service providers can be formulated as an *optimal search problem* [9]. If a provider i is contracted by the center and devotes r_i resources to the task, then it may successfully complete the task with probability $P(r_i)$, providing the agent with value $V - \alpha_i - \beta_i$, or the provider may fail with probability $1 - P(r_i)$, providing the agent with value $-\beta_i$. Weitzman showed that in such a setting, each provider can be characterized by an index given by:

$$z_i = \frac{P(r_i)(V - \alpha_i) - \beta_i}{P(r_i)} \tag{7}$$

The optimal order in which the providers should be procured corresponds to the decreasing order of z_i. That is, the provider with the largest index should be contracted first, followed by the provider with the second largest index, if the first one fails, and so on. Given equations 1 and 2 which specify α_i and β_i, we have $z_i = V - c_i(r_i)/P(r_i)$. Since the cost functions of the providers are non-crossing, and quality of service function, $P(\cdot)$, is the same for all providers, then for a given r, providers with lower cost functions should be procured before those with higher ones.

Theorem 1. *Let M be a set of service providers, $|M| = m$, and assume $c_i(r) \leq c_{i+1}(r)$ for all r. Then when procuring the providers sequentially, i should be contracted before $i + 1$.*

Proof. Proof follows from [9]. $\qquad \square$

Given that we now know the order in which to contract the providers, we can derive the utility of the agent. Assume that there are m providers, and that they are ordered such that $c_i(r) \leq c_{i+1}(r)$ for all r. The expected utility of the agent, given resource-allocation vector $\bar{\mathbf{r}} = (\bar{r}_1, \ldots, \bar{r}_m)$ is:

$$U_A^{\text{seq}}(\bar{\mathbf{r}}) = U_A(\bar{r}_1) + \sum_{i=2}^{m} U_A(\bar{r}_i) \prod_{j=1}^{i-1}(1 - P(\bar{r}_j)) \tag{8}$$

In order to maximize the utility of the agent, set $\nabla U_A^{\text{seq}}(\bar{\mathbf{r}}) = 0$. This results in $m^* = V$ and:

$$\alpha_i^* = V - \left[U_A(\bar{r}_{i+1}) + \sum_{j=i+2}^{m} U_A(\bar{r}_j) \prod_{k=i+1}^{j-1}(1 - P(\bar{r}_k)) \right] \tag{9}$$

for $1 \leq i < m$. Unlike the simultaneous procurement case, α_i^* depends only on the cost functions of providers that are potentially contracted if i fails to complete the task. As was done in the single provider setting, to calculate β_i^*, we define $g_i(r) = c_i'(r)/P'(r)$, and then the optimal resource demand is $\bar{r}_i^* = g_i^{-1}(\alpha_i^*)$. Thus $\beta_i^* = c_i(\bar{r}_i^*) - \alpha_i^* P(\bar{r}_i^*)$.

4   Eliciting Cost Functions

In the previous section we assumed that the cost functions of the providers were known to the agent. Given this information, the agent was able to compute the appropriate values for parameters α_i and β_i so as to incentivise the providers in such a way that the agent's own expected utility was maximised. However, the assumption that the agent has full information is unrealistic in many practical applications. In this section, therefore, we show that it is possible to relax this assumption. In particular, we introduce a family of *task procurement mechanisms* TPM(α, β), such that providers are willing to truthfully reveal their cost functions to the agent, that then uses this information in order to set α_i and β_i appropriately.

Specifically, the agent first decides on a maximum number of providers, m ($1 \leq m \leq n$), it wants to procure services from (the optimal value of m can be determined experimentally, as shown in Section 5) . The agent then executes TPM(α, β), which proceeds as follows:

1. *Cost elicitation:* All providers $i \in \{1, \ldots, n\}$ report their cost functions $\hat{c}_i(\cdot)$ to the agent. We do not assume that providers reveal their true cost functions.

2. *Service provider selection and payment specification:* The agent selects the m providers with the lowest reported cost functions. We denote the set of chosen providers by M. Since we assume that the cost functions are non-crossing, there is no ambiguity in this selection. The agent then calculates α_i and β_i for each provider $i \in M$, and reports these parameters to the providers.

3. *Task execution:* If the agent uses a simultaneous procurement strategy, then all providers in M are asked to perform the task. If the agent uses a sequential procurement strategy then one provider, $i \in M$, is chosen at random to perform the task (Section 4.2 explains why this randomisation is important). If provider i fails, then another provider $j \in M \setminus \{i\}$ is chosen at random. This process continues until either a provider successfully completes the task, or all providers in M have attempted the task once and failed. Note that a provider is always allowed to refuse to attempt the task.

4. *Payment:* Any provider $i \in M$ that was contracted by the agent and successfully completed the task is paid $\alpha_i + \beta_i$. If provider i failed at the task then it receives β_i. All providers in M that were not asked to attempt the task, and those not in M initially, receive zero.

In the rest of this section we describe how to calculate parameters α_i and β_i so that the mechanism is incentive compatible and individually rational.

4.1   Simultaneous Procurement

We start by noting that the two parameters α and β allow us to define a *family* of incentive compatible mechanisms. Then, Theorem 2 characterises the family of such mechanisms for the simultaneous procurement strategy. We let $\hat{c}_{m+1}$ denote the $(m+1)^{th}$ lowest reported cost function, and $r^*(\alpha, c)$ is the optimal investment decision of a provider with a cost function $c(\cdot)$ and when the agent announces parameter α.

Theorem 2. *If, for all $i \in M$, α_i is independent of $\hat{c}_i(\cdot)$, and β_i is given by:*

$$\beta_i = \hat{c}_{m+1}(r^*(\alpha_i, \hat{c}_{m+1})) - VP(r^*(\alpha_i, \hat{c}_{m+1})) \tag{10}$$

then TPM(α, β) with simultaneous procurement is individually rational and incentive compatible.

Proof. Individual rationality holds because a provider i can always refuse the task after learning α_i and β_i. Since all providers in M are asked to attempt the task, the expected utility of $i \in M$ does not depend on the payment or the resources allocated to the task by any other providers $j \in M$. Thus we can look at the incentives of each provider $i \in M$ independently. Consider first provider i with cost function c_i. If $c_i < \hat{c}_{m+1}$ then i has no incentive to announce $\hat{c}_i > \hat{c}_{m+1}$ since by doing so it would guarantee itself a utility of zero. If it revealed its cost function truthfully, then its expected utility is greater than or equal to zero. Now consider the case where provider i has true cost function $c_i > \hat{c}_{m+1}$, but misreports $\hat{c}_i < \hat{c}_{m+1}$ and is selected. Let $U_i(r, c) = \alpha_i P(r) + \beta_i - c(r)$. Note that β_i is set such that $U_i(r^*(\alpha_i, \hat{c}_{m+1}), \hat{c}_{m+1}) = 0$. It follows that $U_i(r^*(\alpha_i, c_i), \hat{c}_{m+1}) < 0$, and since $c_i > \hat{c}_{m+1}$, $U_i(r^*(\alpha_i, c_i), c_i) < U_i(r^*(\alpha_i, c_i), \hat{c}_{m+1}) < 0$. Thus such a provider always receives negative utility. Together with the fact that payments to the providers are independent of their reports, i has no incentive to under-report its cost function.

We now introduce two mechanisms that satisfy the above requirement for incentive compatibility: (1) *uniform pricing* and (2) *discriminatory pricing*. From Theorem 2, since β is given by Equation 10, we only need to worry about setting α_i for all $i \in M$. Now, in the uniform pricing mechanism, $\alpha = \alpha_1 = \ldots = \alpha_m$, and we calculate α by replacing $\bar{r}_j$ in Equation 6 by $r^*(\alpha, \hat{c}_{m+1})$, resulting in following equation:

$$\alpha = V \left[1 - P(r^*(\alpha, \hat{c}_{m+1}))\right]^{m-1} \tag{11}$$

Corollary 1. *Uniform Pricing Mechanism. TPM(α, β) with α and β satisfying Equations 11 and 10 is individually rational and incentive compatible.*

Proof. Since α is independent of any of the reports of providers in M, the proof follows from Theorem 2.

Although uniform pricing is a natural extension of the single-provider case, there are better alternatives. In particular, although from Theorem 2, α_i needs to be independent of $\hat{c}_i$, we can use the cost functions of other providers to calculate α_i. In fact, from Equation 6 the optimal α_i^* in the complete information case can be calculated from $\bar{r}_j, j \neq i$ and thus does not directly depend on $\hat{c}_i$. However, as noted in Section 3.2, if we solve optimally for i as well as all j simultaneously, then α_i^* depends indirectly on $\hat{c}_i$ since $\bar{r}_j^*$ depends on $\bar{r}_i$, and r_i^* in turn depends on $\bar{r}_j^*$. Nevertheless, we can take advantage of the information available about the cost functions of other providers to develop a discriminatory pricing mechanism, though we need to be careful about how we calculate α_i to ensure incentive compatibility. In more detail, we calculate α_i for a specific provider $i \in M$ by solving the following system of equations:

$$\alpha_i = V \prod_{j \in M \setminus \{i\}} \left[1 - P(r^*(\alpha_j, \hat{c}_j))\right] \tag{12}$$

and for all $j \in M \setminus \{i\}$:

$$\alpha_j = V[1 - P(r^*(\alpha_i, \hat{c}_{m+1}))] \prod_{k \in M \setminus \{i,j\}} \left[1 - P(r^*(\alpha_k, \hat{c}_k))\right] \tag{13}$$

Note that we need to derive and solve a separate set of equations for each $i \in M$.

Corollary 2. *Discriminatory Pricing Mechanism. If, for each $i \in M$, α_i is given by independently solving equations 12 and 13, and if β_i is given by Equation 10, TPM(α, β) is incentive compatible and individually rational.*

Proof. Note that $\hat{c}_i$ does not appear in equations 12 and 13, and thus α_i is independent of that provider's report. The proof follows directly from Theorem 2.

4.2 Sequential Procurement

In the full-information setting, described earlier, the order in which the providers are asked to perform the task was shown to be important. However, in the incomplete information setting, we cannot use the reported costs to determine this order since it influences the expected utility of the providers (since those with higher cost functions are less likely to be contracted by the agent). For this reason, the providers are randomly selected in the third stage of the mechanism, and we refer to $i \in M$ as the ith provider in the random sequence, but $\hat{c}_{m+1}$ is the $(m+1)^{th}$ lowest reported cost as before.

Now, whereas in the previous section we were able to use the reported costs of the other providers to adjust the payment without changing the incentives, this is no longer the case in the sequential setting. To see this, note that from Equation 9, α_i^* depends on all providers that appear after i in the sequence. Given these considerations, Theorem 3 reformulates the requirements for incentive compatibility in terms of the sequential procurement setting.

Theorem 3. *If, for all $i \in M$, α_i is independent of any $\hat{c}_j(\cdot)$, $m \geq j \geq i$, and β_i is given by Equation 10, then TPM(α, β) with sequential procurement is individually rational and incentive compatible.*

Proof. Since $\hat{c}_i(\cdot)$ does not affect the payment of providers $j > i$, the quality of service offered by these providers is independent of this report. Furthermore, the position of provider i in the sequence is independent of $\hat{c}_i(\cdot)$. As a result, other than through the allocation decision, provider i's utility is independent of $\hat{c}_i(\cdot)$. Furthermore, analogous to the arguments in Theorem 2, β_i is set such that providers cannot benefit by misreporting in order to change the allocation.

We now present a specific discriminartory payment scheme where $\alpha_m = V$, and α_i for $i < m$ is calculated by modifying Equation 9 to give:

$$\alpha_i = V - \left[U_A(r^*(\alpha_{i+1}, \hat{c}_{m+1})) + \sum_{j=i+2}^{m} U_A(r^*(\alpha_j, \hat{c}_{m+1})) \prod_{k=i+1}^{j-1} [1 - P(r^*(\alpha_k, \hat{c}_{m+1}))] \right]$$

$$(14)$$

Corollary 3. *Sequential Procurement. TPM(α, β) where $\alpha_m = V$, α_i for $i < m$ is given by Equation 14, and β_i is given by Equation 10, is individually rational and incentive compatible.*

Proof. Since α_i is independent of any reports of providers in M, the proof follows directly from Theorem 3.

5   Empirical Evaluation

Having described three incentive compatible and individually rational mechanisms, we now instantiate the quality of service and cost functions, and empirically evaluate our approach. The purpose of this evaluation is two-fold. First, it provides an example of a task allocation domain and how the mechanisms can be applied. Second, it allows us to

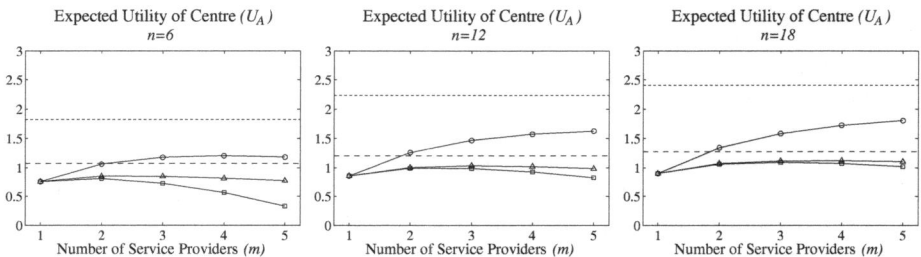

Fig. 2. Simulation results showing the expected utility of the agent for simultaneous procurement with uniform pricing (squares), simultaneous procurement with discriminatory pricing (triangles) and sequential procurement (circles). Results for optimal sequential (short dashed) and simultaneous (long dash) procurement with complete information are also shown.

compare and evaluate the different mechanisms which we presented against the original case in which a single provider is contracted[3]. In more detail, we define the quality of service function as $P(r) = 1/(1 + 1/r)$ and the providers' costs as linear functions $c_i(r) = K_i r$. As required, $P(\cdot)$ is strictly concave, $P(0) = 0$ and $P(r) \to 1$ as $r \to \infty$. Furthermore, note that $c(\cdot)$ is convex (although not strictly convex) and $c(0) = 0$. In the simulations that follow the constants, K_i, are independently drawn from a uniform distribution with support $[1, 2]$.

We choose these functions since they are representative of the general class of functions to which our formalism applies, and also because they yield attractive analytical solutions. For example, each provider maximises its utility by committing resources $r_i^* = \sqrt{\alpha_i/K_i} - 1$ and this will result in a quality of service such that $P(r_i^*) = 1 - \sqrt{K_i/\alpha_i}$. Furthermore, Equation 6 becomes $\alpha_i^* = V \prod_{j \neq i \in M} \sqrt{K_j/\alpha_j^*}$ and can be solved by taking the logarithm of both sides, and using the standard numerical technique of Gauss-Seidel iteration [10], incorporating the voluntary participation constraint.

Figure 2 shows the results of applying the simultaneous (with uniform and discriminatory pricing) and sequential procurement mechanisms in this setting. We vary both the total number of providers, n, and the size of the subset of them that are selected by the agent, m, and in all cases $V = 4$. We first note that the expected utility of the agent when using the simultaneous procurement mechanism with discriminatory pricing always exceeds that of the mechanism with uniform pricing, and thus discriminatory pricing is always preferred. This is not surprising, since discriminatory pricing makes more use of the cost information that is available to the agent. Likewise, the expected utility of the agent when using the sequential procurement mechanism always exceeds that of either simultaneous procurement mechanism, and thus, if sequential procurement is feasible in the specific application domain, then it is always preferred in both the full and incomplete information settings. Interestingly, the difference in utility between the full and incomplete information setting is much larger in the case of

---

[3] Note that our simultaneous and sequential procurement mechanisms are identical in the case that $m = 1$, and since the single provider mechanism is efficient, this is also identical to Matsubara's.

the sequential procurement mechanism compared to either simultaneous procurement mechanism. Finally, the results show that given any specific setting (i.e. the valuation of the agent, the total number of providers and the distribution that describes their cost functions), there is an optimum number of providers to select to procure the task from (either simultaneously or sequentially), and we note that when n is large, procurement from multiple service providers in preferable to procuring from a single one, over a wide range of values of m.

6 Conclusions

In this paper, we considered the problem of procuring computational tasks from self-interested service providers that are able to flexibly manipulate their quality of service in order to maximise their own utility, and we derived a family of task procurement mechanisms that allowed a contracting agent to procure tasks from multiple service providers (either simultaneously or sequentially). Our future work in this area concerns extending these results to the case in which the agent wishes to procure multiple interdependent tasks which exhibit complementary and substitutable valuations. This setting corresponds to the problem of procuring services within a computational workflow, where the entire workflow may be worthless if particular tasks are not completed successfully. Such an extension is likely to require the use of combinatorial auctions, and we are particularly interested in exploring the existence of individually rational and incentive compatible mechanisms that are also optimal.

References

1. Foster, I., Jennings, N.R., Kesselman, C.: Brain Meets Brawn: Why Grid and Agents Need Each Other. In: Proc. of the 3rd Int. Conference on Autonomous Agents and Multiagent Systems, vol. 1, pp. 8–15 (2004)
2. Dash, R.K., Ramchurn, S.D., Jennings, N.R.: Trust-based mechanism design. In: Proc. of the 3rd Int. Conference on Autonomous Agents and Multi-Agent Systems, pp. 748–755 (2004)
3. Porter, R., Ronen, A., Shoham, Y., Tennenholtz, M.: Fault tolerant mechanism design. Artificial Intelligence 172(15), 1783–1799 (2008)
4. Matsubara, S.: Trade of a problem-solving task. In: AAMAS 2003: Proceedings of the second international joint conference on Autonomous agents and multiagent systems, pp. 257–264. ACM, New York (2003)
5. Miller, N., Resnick, P., Zeckhauser, R.: Eliciting informative feedback: The peer-prediction method. Management Science 51(9), 1359 (2005)
6. Papakonstantinou, A., Rogers, A., Gerding, E.H., Jennings, N.R.: A truthful two-stage mechanism for eliciting probabilistic estimates with unknown costs. In: Proc. of the 18th European Conference on Artificial Intelligence, Patras, Greece, pp. 448–452 (2008)
7. Stein, S., Jennings, N.R., Payne, T.: Flexible service provisioning with advance agreements. In: Proc. of the 7th Int. Conference on Autonomous Agents and Multi-Agent Systems, pp. 249–256 (2008)
8. Jehiel, P., Moldovanu, B.: Efficient Design with Interdependent Valuations. Econometrica 69(5), 1237–1259 (2001)
9. Weitzman, M.L.: Optimal search for the best alternative. Econometrica 47(3), 641–654 (1979)
10. Press, W.H., Flannery, B., Teukolsky, S.A., Vetterling, W.T., et al.: Numerical recipes. Cambridge University Press, New York (1986)

An Agent-Oriented Service Model for a Personal Information Manager

Tarek Helmy[1,*], Ali Bahrani[1], and Jeffrey M. Bradshaw[2]

[1] College of Computer Science and Engineering, King Fahd University of Petroleum
and Minerals, Dhahran 31261, Kingdom of Saudi Arabia, Dhahran 31261
{helmy,bahrani}@kfupm.edu.sa
[2] Florida Institute for Human and Machine Cognition (IHMC), Pensacola, FL 32502, USA
jbradshaw@ihmc.us

Abstract. Developing a reusable model of high quality requires consideration
of the full development life cycle of an agent-based system. In this paper, we
discuss an approach based on the Gaia methodology for describing and
designing a service model for a personal information manager based on the
agent-oriented paradigm. The proposed model is shown to be complete, scalable, independent of specific development frameworks, and supportive of a
high degree of autonomous behavior. The extensibility of the model is shown
by elaborating the original model to support speech recognition and calendar
scheduling based on user preferences and learning from history.

Keywords: Agent-oriented software, Personal Information Manager.

1 Introduction

People spend significant time and effort in reading, filtering, searching, and managing
their to-do lists, contacts, e-mail messages, and appointments. Personal Information
Managers (PIM) are meant to reduce the amount of time and number of errors in
management-related activities people perform to complete work-related tasks. One of
the goals of a PIM is to have the right information in the right place, in the right form,
and of sufficient completeness and quality to meet the current need. Hence, scientists
in the fields of computer science and time management have tried (and are still trying)
to increase the productivity of users by inventing new practices and techniques that
help them to better manage their personal information [12, 13].

Some of these practices can be automated or learned. Software applications can be
developed to support such practices and to carry out some actions automatically of
semi-automatically. In addition, a user's preferences can be learned in order to
accomplish personalized actions [11]. Ideally, relationships among applications handling different personal information types are also modeled and leveraged. For example, a new appointment may be created as the result of email exchanges with someone
who is already included in the contacts database.

---

* Tarek Helmy author is on leave from the College of Engineering, Department of Computers
and Automatic Control, Tanta University, Egypt.

R. Kowalczyk et al. (Eds.): SOCASE 2009, LNCS 5907, pp. 24–40, 2009.

The strong dependencies among components of such applications make it difficult to handle modifications in an incremental way. The model described in this paper provides a foundation for a sophisticated PIM that can support a high degree of autonomous behavior as well as the ability to learn. In addition, its extensibility allows people to elaborate the model to support specialized requirements while continuing to take advantage of existing features. Such extensibility is a key to reducing development time and effort.

Our framework-independent PIM model supports the main features of a generic PIM: *Contacts*, *Tasks*, *Calendar* and *Email*. Several artifacts are prepared in different phases of the development life cycle including: *documents*, *schemas*, *tables* and *diagrams*. The Gaia methodology is used to guide us through the development life cycle [8]. The model contains the agent types involved in the system and illustrates how they interact. In addition, the model specifies the responsibilities and permissions of each agent. The paper is organized as follows: Section 2 provides an overview of related work. Section 3 provides details of model development while giving an overview of the Gaia agent-oriented software engineering methodology. Section 4 shows the extensibility of the model through the addition of two new agents. Section 5 presents an execution scenario for the model. Finally, we give our conclusions in Section 6.

2   Related Work

There are a lot of products and tools for managing users' Personal Information (PI). Some of these tools have been proposed in the research domain, whereas others are innovative commercial systems. They can be classified into the following categories, based on the level of integration they provide.

PI-Specific Tools: These tools provide technology aimed at a specific type of PI such as email messages or tasks on a to-do list. In addition, they do not consider integration between distinct PI tools as a primary design goal. EMMA, Towel and Data Mountain are examples of tools in this category.

EMMA is an email system focusing on email management as a process. It manages the sorting, prioritizing, reading, replying, archiving, and deleting of email messages. It caters to a wide variety of users by adopting a knowledge acquisition technique known as "Ripple Down Rules" (RDR). RDR is an incremental technique in which the user starts with an empty knowledge base and adds rules while processing examples [3].

Towel is a task management application that manages users' to-do lists, and was developed by SRI international. It provides a unified environment that enables users to manage their tasks, delegate tasks to others, or collaborate with other users. It uses various AI technologies designed to save user's time and load, and to improve task performance [10].

Data Mountain is a technique that allows users to place documents at arbitrary positions in a 3D desktop virtual environment using a simple 2D interaction technique. Its interface is designed specifically to take the advantage of human spatial memory in managing documents (i.e. the ability to remember where you put something). It is designed with a fixed viewpoint so that users need not to navigate around the space. Users can identify and distinguish between documents by relying on thumbnail representations and pop-up titles [7].

Systems providing integration between distinct PI-specific tools: Tools in this category offer limited integration based on some kinds of structured information. For example, a tool in this category may allow the user to access the contact manager when selecting an email address in a message. Stuff-I've-Seen (SIS) is an example of a tool in this category. It has been developed to make it easy for people to find information they have seen before. There are two main concepts in the design of SIS that help it to achieve information reuse. First, the system provides a unified index of information that a person has seen on his computer. Second, because a person has seen the information before, rich contextual cues such as time, author, thumbnails, and previews can be used to search for and present information. SIS indexes many types of information such as emails, web pages, documents, media files, calendar appointments, file system hierarchy, email folder hierarchies, favorites, and web pages history. All of these are integrated into a single index, regardless of the form or origin of the information [2].

Systems embedding additional support for managing multiple types of information within one PI-specific tool: TaskMaster is an example of tools in this category. It allows the user to manage multiple types of PI with one PIM tool through the embedding of extra functionality. It is a client that provides a mechanism for labeling any item of information with to-do metadata. In addition, it manages multiple types of PI [1].

Systems consolidating the management of all PI in a single new interface: This is in contrast to systems that embed PI within an existing tool. Examples of tools in this category include MEMOIRS, ContactMap and UMEA, which unify PI management in a single interface based on time, contacts, and activities [1]. **A MEMOIR is** a prototype that has been developed based on the chronological organization of the PI. It is based on integrating a user's diary and filing the system with a chronological mechanism. The prototype is designed to enable PI retrieval based on temporal context, and to promote retrieving by recognition rather than by coloring items [1]. **ContactMap** integrates the management of different PI based on the representation of a user's social network, which is derived from the user's address book. The interface maps between a social network, files, bookmarks and email messages. It also enables users to use the information retrieved from the address book for communication. Thus, **ContactMap** integrates both information management and communication functionality [1]. The **UMEA** allows the user to organize multiple types of PI based on their projects. The design rationale is based on the observation that a user's activity often involves multiple PIM tools. Hence, the user is asked to provide the project name that he is currently working on. Thereafter, all the information details that is associated with the proposed model [1].

Our model differs from the above-mentioned products in several respects. First, it manages all major PI types, i.e.: Tasks, Contacts, Calendar and Email. By way of contrast, all of the above-mentioned products manage only a few PI types and ignore the relationships among them. Second, our model is agent-oriented: it uses the features of the agent as an abstraction unit that is proactive, reactive and able to communicate with other agents. Finally, our model is extensible: it allows the designer to add new agents and incorporate their functionalities to the overall model.

3 Building the Model

3.1 Rationale for Selecting Gaia

Formal guidelines on how to progress through different phases of the development life cycle are ultimately intended to save time and effort. Effective guidelines of this sort are designed for usability, built according to best practices, and describe the important steps that the designer should follow [4]. In general, a software development methodology consists of process, heuristic rules, artifacts, notations and pattern [4, 5]. A process is a sequence of phases and activities that guide the developer to build a system. Heuristic rules are those supporting the developer in making relevant choices. Artifacts are diagrams, schemas or documents in the form of text or graphics. Notations are those which used in representing the artifacts. A pattern is what can be applied to address issues that arise in common situations [4, 5].

Over the past two decades, complex systems have been engineered using powerful and natural high-level abstractions. Examples of such abstractions include procedural abstractions, abstract data types, objects, and components. Software agents can be seen as advanced abstractions that may be used by software developers to more naturally understand, develop, and model complex systems. Since agents provide an advanced abstraction for complex systems, it is helpful to use software engineering techniques that are specifically tailored for them. Existing software development techniques such as object-oriented analysis and design fail to represent agents' characteristics. They fail to adequately capture agents' flexible and autonomous behaviors, the richness of the agents' interactions, and the complexity of the agents' system organization structure [9].

Gaia has become well-known as an agent-oriented software engineering methodology because it is tailored specifically to the analysis and design of agent-based systems and deals with macro-level (social) as well as micro-level (agent) design aspects [8]. In addition, it captures agents' characteristics of being proactive, reactive, and having social ability. The Gaia methodology is characterized by its:

1. *Precision:* the live-ness and safety properties in role definition make it accurate and prevent misunderstanding of the modeled functionality.
2. *Accessibility:* Gaia is easy to understand and use due to its simple models and clarity.
3. *Expressiveness:* Gaia can handle a large variety of systems due to its generic structure.
4. *Modularity:* Gaia is modular because of its building blocks such as roles, protocols and activities.

The Gaia methodology enables analysts to move from abstract concepts to increasingly concrete ones. Abstract concepts are those used during analysis to conceptualize the system like roles, permissions, responsibilities, protocols, activities, live-ness, and safety properties. However, concrete concepts are those that are used within the design process and have direct counterparts in the runtime system like agent types and acquaintances. The following sections define the different phases we went through in order to build the model and the artifacts produced in each phase.

3.2 Requirements

Since the Gaia methodology has two phases only (analysis and design), requirements capture should be done independently beforehand. The most popular way to capture the potential function requirements of the system is to employ 'use cases' where each use case represents one or more scenarios that demonstrate how the system should interact with users or other systems. Use cases are not specific to any particular software development methodology; hence they can be used to capture the functional requirements of multi-agent systems without modification [4]. The first step in developing the use cases is to define the set of actors that are involved in the story, where actors are the different people or devices that use the system [6]. The only actor in our PIM example is the user who is accessing the system and is willing to manage his/her personal information. The second step is identifying the system's major use cases.

In the proposed PIM, the major use cases identified are: '*Maintain Tasks*', '*Maintain Contacts*', '*Maintain Calendar*' and '*Maintain Emails.*' In addition to these four main use cases, there is one use case '*Verify User*' which needs to be executed as a prerequisite for each of the main use cases. Hence, this use case is related to the main use cases and the relationship is of type <<includes>>. See Figure 1 below for the main use cases diagram.

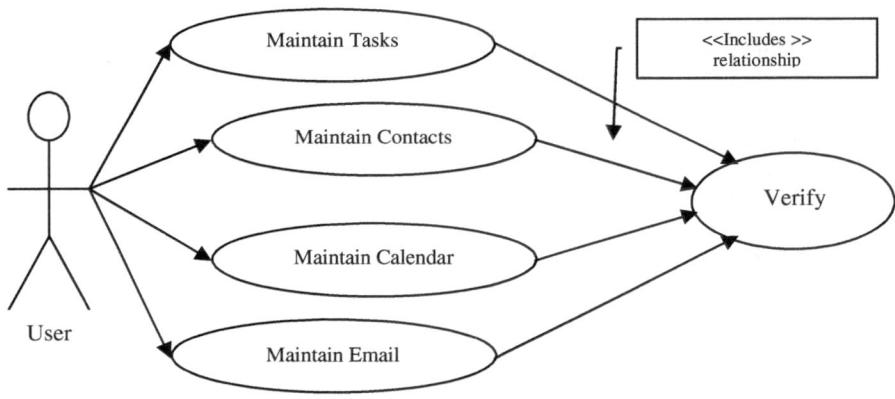

Fig. 1. Main Use Cases Diagram

3.3 Analysis

The objective of the analysis phase in Gaia methodology is to develop an understanding of the system and its structure. It includes the identifications of the environment model, the role model and the interaction model of the system. The following sections discuss these models for the proposed PIM.

3.3.1 The Environment Model
The environment is treated in Gaia in terms of abstract computational resources such as variables that are made available to agents for sensing, effecting, or consuming. Such variables are accessed by agents to be read or changed, or to access their values.

Hence, the environment model can be shown as a list of abstract computational resources associated with symbolic names and characterized by the types of actions agents are performing on them. In addition, it is also possible to add textual comments and descriptions to each resource.

Our PIM model uses four resources:, *Tasks Database, Contacts Database, Calendar Database*, and *Emails Database*. Users interacting with the system are able to manage their personal information using these resources. Here the PI is categorized by the user so that only her/his information can be accessed. To enforce this requirement, a name and password will be required at system startup. Obviously, the system itself also needs read/write access to these resources.

Throughout the analysis, design and implementation phases of the system, the tasks, contacts, calendar items and emails resources are represented in two-dimensional arrays. The first index of the array represents the user identification, whereas the second index represents the resource's element identification. For example, the representation Task[i][j] means the jth task for the user i. These discussions are summarized in Figure 2, which shows the environment model of the system.

- *Change Tasks[i][j] where i=1,...total_users, and j=1,... total_tasks.*

- *Change Contacts[i][j] where i=1,...total_users, and j=1,... total_contacts.*

- *Change Calendar Items [i][j] where i=1,... total_users, and j=1, ...total_calendar_items.*

- *Change Emails[i][j] where i=1,...total_users, and j=1,... total_emails.*

Fig. 2. The Environment Model of the System

3.3.2 The Role Model

Defining a multi-agent system using roles is quite a natural way of thinking [8, 14]. For example, if we consider a typical company as a human organization, we can define several roles such as 'president', 'vice-president', 'manager' and so on. These roles will be instantiated with actual individuals such that an individual can take the role of a 'president' and another one can take the role of 'vice-president'. It is also possible in some small companies that an individual can take more than one role. For example, an individual can take the role of being a 'mail fetcher' and 'office cleaner'. In addition, it is possible that a role can be instantiated to more than one individual such as the 'salesman' role. The role model in Gaia methodology identifies the key roles in the system and their permissions and responsibilities attributes [9]. Permission attributes identify the resources that can be used to carry out the role and resource's limits ('read', 'change' or 'generate'). In order to represent permissions, Gaia makes use of the same notation already used for representing the environment resources. However, the attributes associated with resources are no longer representing what can be done with such resources from the perspective of the environment. On the other hand, the attributes associated with resources are representing what the agent's role must, and must not, be allowed to do in order to accomplish the role's requirements [14]. Conversely, the responsibilities' attributes are those that determine

Table 1. System's Roles List

Personal Information Type	Roles
Task	- Task Creator, Task Modifier - Task Deleter, Task Reassigner
Contact	- Contact Creator, Contact Modifier - Contact Deleter, Contact Searcher - Contact Sender
Calendar Items	- Calendar Creator, Calendar Modifier - Calendar Deleter, Calendar Reminder - Calendar Inviter, Calendar Invitation Tracker
Emails	- Email Creator, Email Deleter - Email Sender, Email Receiver

the expected behavior of a role or, in other words, its *obligations* [14]. Responsibilities are divided into two types: live-ness responsibilities and safety responsibilities. Live-ness responsibilities are those that state "something good happens." They are so called because they tend to say that the agent carrying out the role is still alive. They are specified via a live-ness expression that defines the lifecycle of the role.

The general form of live-ness expression is as follows: Role_Name = expression, where "Role_Name" is the name of the role whose live-ness responsibilities are being defined, and "expression" is the live-ness expression defining the live-ness properties of Role_Name. The atomic components of a live-ness expression are either activities or protocols. An activity is somewhat like a method in object-oriented terms or a procedure in procedural programming language. An activity is a unit of action that the agent may perform that does not involve interaction with any other agent. On the other hand, protocols are activities that require interaction with other agents. An activity is visually distinguished from a protocol through underlining. Sometimes, it is insufficient to specify the live-ness responsibilities of a role. This is because an agent carrying out a role will be required to maintain certain invariants while executing. For example, we might require that a particular agent taking part in an e-commerce application never spends more money than it has allocated. These invariants are called 'safety conditions,' because they usually relate to the absence of some undesirable condition arising. Safety conditions in Gaia are specified by means of a list of predicates. The list is represented as a bulleted list such that each item in the list expresses individual safety responsibility.

The proposed PIM manages four types of PI namely: *Tasks, Contacts, Calendar items* and *Emails*. In order to satisfy the requirements, we have identified the required roles for each PI type, and these are listed in Table 1. In addition to the roles identified in the Table 1, we need these additional roles: *Facilitator, Task, Contact, Calendar* and *Email*. The facilitator is used to perform communication services such as forwarding messages and displaying interface screens. With this in mind, the task, contact, calendar and email roles act as a middle layer between the facilitator role and the other roles in the table. In other words, the roles are tiered into three levels. The

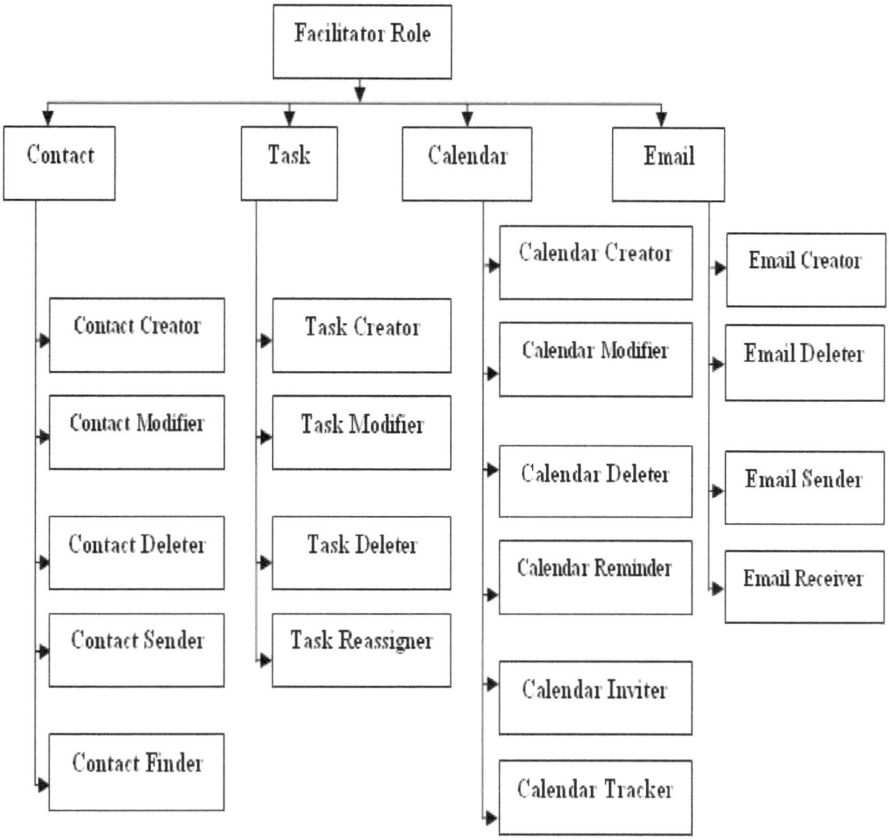

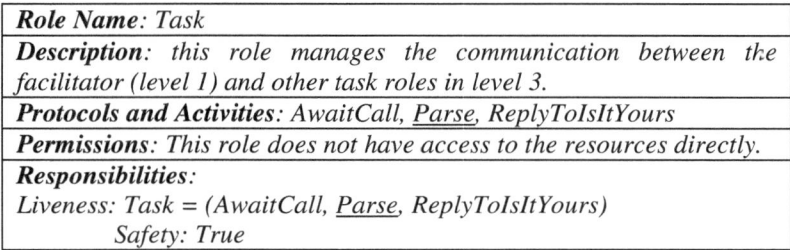

Fig. 3. Roles Hierarchy

first level contains only the facilitator role. The second level contains the task, contact, calendar and emails roles while all the other roles are in the third level. When the facilitator receives a command from the user, it forwards it to the 2nd level roles asking them whether they can satisfy the command. Then, they parse the command and reply with the name of the responsible 3rd level role that can execute the command.

Role Name: Task
Description: this role manages the communication between the facilitator (level 1) and other task roles in level 3.
Protocols and Activities: AwaitCall, _Parse_, ReplyToIsItYours
Permissions: This role does not have access to the resources directly.
Responsibilities: Liveness: Task = (AwaitCall, _Parse_, ReplyToIsItYours)   Safety: True

Fig. 4. Schema for the Task Role

Role Name: TaskCreator
Description: Create a new Task Item for the logged on user.
Protocols and Activities: AwaitCall, CreateTask
Permissions: *Reads supplied LoggedUser //Read all tasks for the user who logged on* *TaskItems[LoggedUser] //Create a new TaskItem i for the user who logged on* *Generates TaskItems[LoggedUser][i].*
Responsibilities: *Liveness: TaskCreator = (AwaitCall.CreateTask)*   *Safety: True*

Fig. 5. Schema for the Task Creator Role

Then, the facilitator forwards the command to that specific responsible 3rd level agent asking for execution. Figure 3 shows the tiers of the model. After we have identified the role hierarchy, we have defined all roles mentioned in the hierarchy using the recommended Gaia's role schema. Figure 4 and 5 show the schemas for the *'Task'* and *'TaskCreator'* roles as examples.

3.3.3 The Interaction Model

This model captures the dependencies and relationships between the roles of the system. Each interaction between two roles has a protocol definition. A protocol definition consists of these attributes:

- *Protocol Name*: brief description that captures the nature of the interaction.
- *Initiator*: the role(s) responsible for starting the interaction.
- *Partner*: the responder role(s) with which the initiator interacts.
- *Inputs*: information supplied to the protocol responder during interaction.
- *Description*: textual description explaining the purpose of the protocol and the processing activities implied in its execution.

In order to build the interaction model of the proposed PIM, we have defined each interaction (protocol) named in the role model using Gaia's protocol definition template. For example, in the Task role schema defined in Figure 4, we have two protocols *'AwaitCall'* and *'ReplyToIsItYours'* which are defined in Figures 6 and 7 respectively.

Role Name: Task		
Protocol Name: AwaitCall		
Initiator:	*Partner:*	*Input:*
Facilitator	*Task*	*User's command*
Description:		*Output:*
When the user commands the system to do something, the facilitator will send the command to all 2nd level roles (including task role) asking them if the command is related to them or not.		*Command will be parsed*

Fig. 6. Definition of AwaitCall Protocol

Role Name: Task		
Protocol Name: ReplyToIsItYours		
Initiator:	Partner:	Input:
Task	Facilitator	None
Description:		Output:
This interaction occurs as a reply to the facilitator in his request about if the command is related to this role or not. If it is related then a responsible role from level 3 will be sent as output.		Yes/No, and the responsible agent from level 3.

Fig. 7. Definition of ReplyToIsItYours Protocol

3.4 Design

The objective of the design phase is to transform the abstract models derived during the analysis phase into a sufficiently low level of abstractions in order to implement agents. The design phase includes the generation of the agent model and the acquaintance model. The following sections discuss these models for our PIM.

3.4.1 The Agent Model

The purpose of the agent model is to document the various agent types that will be used in the system and the agent instances that will realize these agent types at runtime. An agent type is a set of agent roles. However, sometimes there is a one-to-one correspondence between roles and agent types. A designer can choose to package a number of closely related roles in the same agent type for the purpose of convenience or efficiency. The designer may want to optimize the design by aggregating a number of agent roles into a single type that carries out all the functionalities required by all roles. Later, only this agent type needs to be delivered. In our case there is a one-to-one correspondence between roles and agent types. That means we will deliver as many agent types as roles defined in Figure 3.

3.4.2 The Acquaintance Model

This model defines the communication links that exist between agent types and simply indicates that communication exists. However, it does not define what messages are sent or when they should be sent. The purpose of this model is to identify any potential communication bottlenecks that may cause problems at runtime. The acquaintance model is a directed graph with nodes in the graph corresponding to agent types and arcs corresponding to communication pathways. Figure 8 below shows the acquaintance model of the system.

3.5 Prototype Implementation

We have developed a prototype for the proposed PIM model. The prototype was developed using the Microsoft SQL Server 2005 Express Edition as a back-end database and the AgentBuilder as a multi-agent development tool. Using Gaia methodology in the analysis and design phases helps us developing the prototype in minimal effort. The artifacts produced in the analysis and design phases can be directly transformed to code. For example, if we consider the *TaskCreator* role produced in the analysis phase and defined in Figure 5, we find that the live-ness expression is: TaskCreator = (AwaitCall · CreateTask), where AwaitCall is a protocol involving communication with another agent, and <u>CreateTask</u> is an activity that the agent can do without any communication.

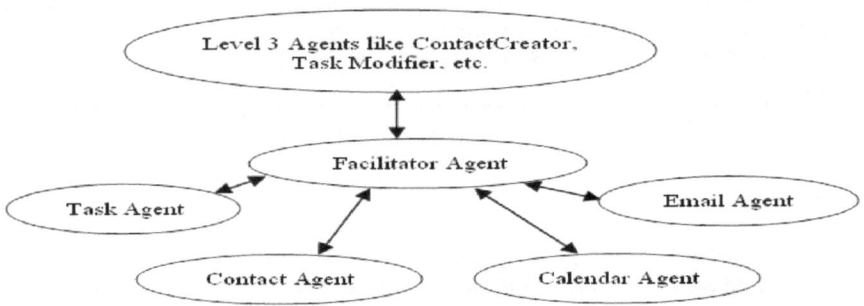

Fig. 8. Acquaintance Model

Similarly, in order to find the definition of the AwaitCall protocol, we refer to the interaction model developed also in the analysis phase [Figure 9]. The definition of the protocol specifies that the initiator for the communication is the facilitator agent, while the TaskCreator agent is the partner. It also specifies that the input to that interaction is the Task details, and the output expected after the interaction is inserting the task into the database table. This discussion is directly mapped into the agent's rules in the AgentBuilder tool as shown in Figure 10.

Role Name: TaskCreator			
Protocol Name: AwaitCall			
Initiator:	*Partner:*		*Input:*
Facilitator	*TaskCreator*		*Task Details*
Description:			*Output:*
In this interaction, the Facilitator agent sends the task details to the TaskCreator in order to create a new task in the database.			*New task created*

Fig. 9. Definition of AwaitCall Protocol

The interpretation of the snippet in Figure 10 is as follows. If the KQML message received is from the "*Facilitator*" (line 1 in WHEN section) and asks the receiver (line 2 in WHEN section) to execute the command embedded into the message (line 3 in WHEN section), then the receiver executes a CreateTask method of object $my-Task (line 3 in THEN section). Line 4 (in the THEN section) is used to print to the agent's console.

> **WHEN:**
> 1. (%message.sender EQUALS "Facilitator")
> 2. (%message.performative EQUALS "ask-one")
> 3. (%message.replyWith EQUALS "doIt")
> **THEN:**
> 1. DO SystemOutPrintln ("TaskCreator Agent is going to create the task")
> 2. SET_TEMPORARY $myTask TO %message.content.myTask
> 3. DO CreationReturnedValue = $myTask.CreateTask (%message.content)
> 4. DO SystemOutPrintln ("Task is created successfully")

Fig. 10. Behavioral Rules of TaskCreatorAgent

4 Model Extensibility

In addition to benefiting from agents' characteristics, a primary objective of our approach is to build a PIM model that can be extended by researchers in order to test their new practices and algorithms. In this section, we study the ability of the model to scale up by adding two relatively sophisticated agents: the *"Speech Recognition Agent"* and the *"Calendar Scheduler Agent"*.

4.1 A Speech Recognition Agent

The purpose of the speech recognition agent is to transform speech into text. The speech recognition agent does not have direct access to the system resources. Instead,

Role Name: SpeechRecognition
Description: This role recognizes speech and transfers it into text
Protocols and Activities: *ListenToCommand*, *ConvertToText*, *SendTextCommand*.
Permissions: //This agent does not have access to the resources directly
Responsibilities: Liveness: SpeechRecognition = (*ListenToCommand* . *ConvertToText* . SendTextCommand)   Safety: * True

Fig. 11. Schema for the role SpeechRecognition

Role Name: SpeechRecognition		
Protocol Name: SendTextCommand		
Initiator:	Partner:	Input:
SpeechRecognition	Facilitator	Command as text
Description:		Output:
This interaction occurs to transfer the command to the facilitator after it is converted to text		Facilitator will continue the execution process of the command

Fig. 12. Definition of SendTextCommand Protocol

it sends the command to the facilitator agent in the form of text. Figure 11 shows the role's schema while Figure 12 shows the definition of the unique protocol involved: *'SendTextCommand'*.

4.2   A Calendar Scheduling Agent

This agent schedules the meetings of users on their behalf based on their preferences and the usage history. It also negotiates timing with other users until agreement is reached. Users' preferences are stored as profiles in the database. The user needs to answer a number of questions posed by the agent in order to allow supervisory learning. Later, the agent will consult the user's profile to schedule meetings at the user's preferred times. Here are some questions that can be stored in the user's profile:

1-   Do you like your meetings to be contiguous or scattered throughout the day?
2-   If you like your meetings to be scattered, what is the minimum time needed in between?
3-   What are your block days of the week (i.e. the days that you do not want to have meetings in)?
4-   What are your block times of the day (i.e. the times that you do not want the agent to schedule meetings during)?
5-   What is the maximum number of meetings per day?
6-   Who are the people that have high priority? Meetings with high priority people are given higher precedence in scheduling. For example, meetings with the user's manager cannot be negotiated. In addition, in case of conflict, meetings with low priority people will be rescheduled to another time.
7-   What is the maximum number of negotiations allowed per meeting without involving the user? This question is asked in order to stop the agent from negotiating meeting times infinitely. When negotiation exceeds a user-defined limit, the user is able to review his schedules manually and negotiate.

In addition to this supervisory learning, the agent learns the user's preferences from the history. Clearly, this agent needs some functionality provided by the model. It needs to access and update the user's calendar and contacts, as well as, it needs the emailing functionality to negotiate meeting times. Hence, the model is going to be extended. Considering the meeting scheduling problem, the user can be either an organizer or an attendee. Being an organizer means s/he needs to invite other people, whereas an attendee is invited to a meeting. If the user is an organizer, then the scheduling and negotiation process will be as follows:

a.   The user asks the agent to schedule a meeting with someone.
b.   The agent consults the user's preferences and sends an invitation to other members by using the emailing functionality provided by the model.
c.   The agent receives *"Accept,* or *Reject* or *Propose New Time"* replies from the attendees.
d.   Steps (b) and (c) are repeated until satisfaction.

On the other hand, if the user is an attendee then the scheduling and negotiation process will be as follows:

a. The user receives a meeting invitation by email.
b. The agent consults the user's preferences and sends *"Accept,* or *Reject,* or *Propose New Time"* reply to the organizer.
c. Steps (a) and (b) are repeated until satisfaction.

Hence, we have two roles for the CalendarScheduler agent: organizer and attendee. Figure 13 shows the role's schemas for the organizer role while Figure 14 and 15 show the definitions of the protocols involved in its live-ness expression.

Role Name: *Organizer*
Description: *This role organizes a meeting and sends invitations to attendees.*
Protocols and Activities: <u>*ScheduleIt*</u>, *InviteAttendees, ReceiveAcceptReject*
Permissions: *//This agent does not have access to the resources directly*
Responsibilities:
Liveness: Organizer = (<u>ScheduleIt</u> . InviteAttendees . ReceiveAcceptReject)
Safety: True

Fig. 13. Schema for the role Organizer

Role Name: *Organizer*		
Protocol Name: *InviteAttendees*		
Initiator:	Partner:	Input:
Organizer	EmailSender	Meeting Details
Description:		Output:
Send an invitation to attendees.		Invitation will be sent

Fig. 14. Definition of InviteAttendees Protocol

Role Name: *Organizer*		
Protocol Name: *ReceiveAcceptReject*		
Initiator:	Partner:	Input:
EmailReceiver	Organizer	Response
Description:		Output:
Interaction occurs to deliver the invitation response to the organizer		

Fig. 15. Definition of ReceiveAcceptReject Protocol

5 Execution Scenario

Assume that a user wants to create a new task and insert it into the database. The following steps are executed:

1. When users start the prototype, the facilitator agent does not have any belief about their identity. Hence, the facilitator will ask for a name and password. If they are valid, then the facilitator will add this fact to its beliefs and redirect the user to the next step. Otherwise, s/he will be asked to enter the name and password again.

2. The facilitator agent displays a command frame to allow a user to type a command. Assume the user enters the command: 'Can you please create a task "Work on initial conference paper" starting 09/04/2008 and due date is 10/04/2008 with high importance and 30% completed'. Figure 16 shows a snapshot of the command.

3. When the user clicks on the Execute button, the facilitator agent will forward the command as it is to the second-level agents in the hierarchy of the agent model, asking them whether the command is related to any one of them.

4. Each second-level agent will study the command and check whether it is relevant to itself or not. The result of this checking will be sent to the facilitator with the name of the responsible agent in third-level agents. In this scenario, the task agent will reply by saying the command is mine and the specific responsible third-level agent is the "*TaskCreator*" agent [Figure 17]. Other second-level agents will reply by saying that the request is not related to them. In addition, the task agent parses the command and understands it.

5. The facilitator displays a task confirmation screen based on how the task agent has interpreted the command [Figure 18].

6. The user can accept the system's interpretation or change any values. Once the user clicks on the Save button, the facilitator agent will send the command to the 3rd responsible agent to execute it, which in our case is the "*TaskCreator*".

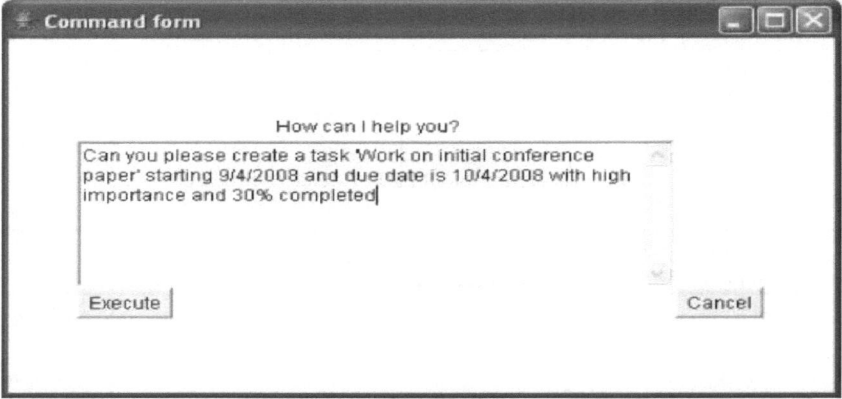

Fig. 16. Snapshot of User's Command in the Scenario

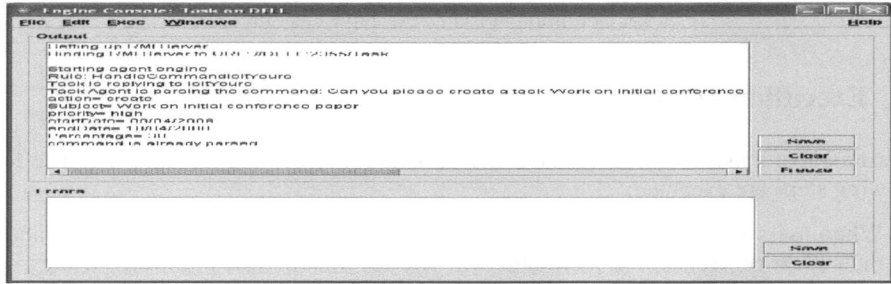

Fig. 17. Snapshot of the TaskAgent Output after Parsing the Command

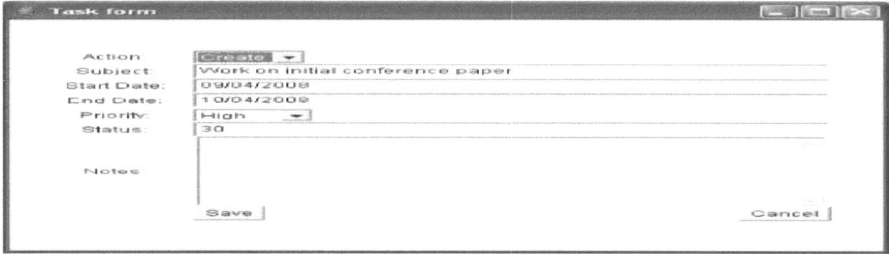

Fig. 18. Snapshot for the Task Confirmation and Editing Screen

6   Conclusion and Future Work

In this paper, we have discussed the importance of following an agent-oriented soft-ware engineering methodology to build a high quality PIM model. The paper provides all artifacts needed by an agent developer in order to implement an agent-based PIM. We have defined all roles involved in the system and their interaction protocols. The proposed model can be used in conjunction with any development framework. Indeed, in our own work, we decided to use AgentBuilder in the prototype only after we had completed and produced all the requirements, analysis, and design artifacts. In the model, we have enabled agents to express one of their most important characteristics, namely, autonomy. Agents in the model can communicate and cooperate with as much or as little user interaction is necessary to do the job. The proposed model represents a foundation for sophisticated PIMs that have high degree of autonomous behavior and ability to learn. In addition, we demonstrate the property of extensibility. We were able to add two agents and incorporate their functionalities to the overall model. With only a few new protocol definitions, the new agents were able to plug in and cooperate with others. Other possible future work includes replicating the model using other software engineering methodologies in order to explore the advantages and disadvantages of each one.

Acknowledgments

We would like to thank King Fahd University of Petroleum and Minerals for support-ing this research work and providing the computing facilities.

References

[1]   Boardman, R.: Improving Tool Support for Personal Information Management. Thesis Report for the degree of Doctor of Philosophy, pp. 30–60 (2004)

[2]   Dumais, S., Cutrell, E., Cadiz, J., Jancke, G., Sarin, R., Robbins, D.C.: Stuff I've seen: a system for personal information retrieval and re-use. In: Proceedings of the 26th Annual international ACM SIGIR Conference on Research and Development in information Re-trieval, Canada (2003), http://doi.acm.org/10.1145/860435.860451

[3] Mao, X., Wang, J., Chen, J.: Modeling Organization Structure of Multi-Agent System. In: Proceedings of the IEEE/WIC/ACM international Conference on Intelligent Agent Technology, pp. 116–119 (2005),
http://dx.doi.org/10.1109/IAT.2005.102

[4] Nikraz, M., Caire, G., Bahri, P.A.: A methodology for the analysis and design of multi-agent systems using JADE. Computer Systems Science and Engineering 21(2) (2006)

[5] Luck, M., Ashri, R., d'Inverno, M.: Agent-Based Software Development. Artech House, Inc. (2004)

[6] Pressman, R.S.: Software Engineering: a Practitioner's Approach. McGraw-Hill Science/Engineering/Math, New York (2004)

[7] Robertson, G., Czerwinski, M., Larson, K., Robbins, D.C., Thiel, D., van Dantzich, M.: Data Mountain: using spatial memory for document management. In: Proceedings of the 11th Annual ACM Symposium on User interface Software and Technology, USA, November 01 - 04, pp. 153–162 (1998),
http://doi.acm.org/10.1145/288392.288596

[8] Wooldridge, M., Jennings, N.R., Kinny, D.: The Gaia Methodology for Agent-Oriented Analysis and Design. In: Autonomous Agents and Multi-Agent Systems, pp. 285–312 (2000), http://dx.doi.org/10.1023/A:1010071910869

[9] Zambonelli, F., Jennings, N.R., Wooldridge, M.: Developing multi-agent systems: The Gaia methodology. ACM Trans. Software Engineering. Method, 317–370 (2003),
http://doi.acm.org/10.1145/958961.958963

[10] Myers, K., Berry, P., Blythe, J., Conley, K., Gervasio, M., McGuinness, D., Morley, D., Pfeffer, A., Pollack, M., Tambe, M.: An Intelligent Personal Assistant for Task and Time Management. AI Magazine 28(2), 47–61 (2007)

[11] Helmy, T.: Towards a User-Centric Web Portals Management. International Journal of Information Technology 12(1), 1–15 (2006)

[12] Jones, W.: Keeping Found Things Found: The Study and Practice of Personal Information Management. Morgan Kaufmann Publishers, Burlington (2008)

[13] Jones, W., Teevan, J. (eds.): Personal Information Management. University of Washington Press, Seattle (2007)

[14] Bradshaw, J.M., Feltovich, P.J., Johnson, M., Bunch, L., Breedy, M., Jung, H., Lott, J., Uszok, A.: Coordination in human-agent-robot teamwork. In: Proceedings of the 2008 International Symposium on Collaborative Technologies and Systems (CTS 2008), Special Session on Collaborative Robots and Human Robot Interaction, Irvine, CA, May 19-23 (2008), pp. 467–476 (2008)

Agent-Based Context Consistency Management in Smart Space Environments*

Wan-rong Jih, Jane Yung-jen Hsu, and Han-Wen Chang

Department of Computer Science and Information Engineering
National Taiwan University
jih@agents.csie.ntu.edu.tw, yjhsu@csie.ntu.edu.tw, r96922005@ntu.edu.tw

Abstract. Context-aware systems in smart space environments must be aware of the context of their surroundings and adapt to changes in highly dynamic environments. Data management of contextual information is different from traditional approaches because the contextual information is dynamic, transient, and fallible in nature. Consequently, the capability to detect context inconsistency and maintain consistent contextual information are two key issues for context management. We propose an ontology-based model for representing, deducing, and managing consistent contextual information. In addition, we use ontology reasoning to detect and resolve context inconsistency problems, which will be described in a *Smart Alarm Clock* scenario.

1 Introduction

In the past several years mobile devices, such as smart phone, personal digital assistants (PDAs), and wireless sensors, have been increasingly popular. Moreover, many tiny, battery-powered, and wireless-enabled devices have been deployed by researchers in smart spaces with the goal of collecting contextual information about their residents. Customized information can be delivered across mobile devices, based on specific contexts (location, time, environment, *etc.*) of the user. The Aware Home[1], Place Lab[2], Smart Meeting Room[3],and vehicles[4] provide intelligent and adaptive service environments for assisting users in concentrating on their specific tasks.

Context-awareness is the essential characteristic of a smart space, and using technology to achieve context-awareness is a type of intelligent computing. Within a richly equipped and networked environment, users need not carry any devices on their person; instead, applications adapt the available resources to deliver services to users in the vicinity, as well as track the location of these users. Cyberguide[5] uses the user's locations to provide an interactive map service. In the Active Badge[6], every user wears a small infrared device, which generates a

---

* This work was partially supported by grants from the National Science Council, Taiwan (NSC 96-2628-E-002-173-MY3), Excellent Research Projects of National Taiwan University (97R0062-06), and Intel Corporation.

R. Kowalczyk et al. (Eds.): SOCASE 2009, LNCS 5907, pp. 41–55, 2009.

unique signal and can be used to identify the user. Xerox PARCTab[7] is a personal digital assistant that uses an infrared cellular network for communication. Bat Teleporting[8] is an ultrasound indoor location system.

In a smart space, augmented appliances, stationary computers, and mobile sensors can be used to capture raw contextual information (*e.g.* temperature, spatial data, network measurement, and environmental factors), and consequently a context-aware system needs to understand the meaning of a given context. Therefore, developing a model to represent contextual information is the first step of designing context-aware systems. Context-aware services require the high-level description about various user states and environmental situations. However, high-level context cannot be directly acquired from sensors. The capability to infer high-level contexts from existing knowledge is required in context-aware systems. Consequently, how to derive high-level contexts is the second step in designing context-aware systems. As people may move within and between environments at any time, it is increasingly important that computers adapt to context information in order to appropriately respond to the needs of users. How to deliver the right services to the right places at the right time is then the third step in designing context-aware systems. Inconsistent contexts may appear in context-aware systems as systems react to the rapid change of contextual information. Systems having inconsistent knowledge will fail to provide correct services. Therefore, a context-aware system must maintain a consistency knowledge base and react to the dynamic change of contexts, which is the fourth step in designing context-aware systems.

In this research, we leverage multi-agent and semantic web technologies that provide the means to express context and use abstract representations to derive usable context for proactively delivering context-aware service to the user. We propose an ontology-base model for supporting context management, which can provide high-level context reasoning and detect knowledge inconsistency. In addition, a *Smart Alarm Clock* scenario serves as an explanatory aid in illuminating the details of our research.

2   Background Technologies

An overview of the context models, context reasoning, and ontology is introduced in this section.

2.1   Context Representation

Context is mainly characterized by four dimensions[9]: location, identity, activity and time. Location refers to the exact position of the user. If we know a person's identity, we could easily identify related information such as birth date, social connectivity, or email addresses when the appropriate access to user-related data is given. Knowing the location of an entity, we could determine its nearby objects and users.

Many context-aware systems concentrate on location aware services. Ye *et al.*[10] use a lattice model to represent spatial structure, which can deal with syntactic and semantic labels. This general spatial model provides both absolute and relative references for geographic positions, as well as both the containment and connection relationships. MINDSWAP Group at University of Maryland Institute for Advanced Computer Studies developed Semantic geoStuff[1] to express basic geographic features such as countries, cities, and relationships between these spatial descriptors.

The RFC 2445[2], which defines the iCalendar format for calendaring and scheduling applications, enables users to create personal event data. Google Calendar[3] is a popular web-based calendar which supports the iCalendar standard and allows users to share their own personal activities with others. These human activities can be broken down into people, time, and location. Consequently, the contents of a person's schedule can help us to derive his or her location at a given time.

2.2 Ontology

Strang and Linnhoff-popien[11] concluded that an ontology is the most expressive model. Gruber[12] defines ontology as an "explicit specification of a conceptualization". An ontology is developed to capture the conceptual understanding of a domain in a generic way and provides a semantic basis for grounding fine-grained knowledge.

COBRA-ONT[13] provides key requirements for modeling context in smart meeting applications. It defines concepts and relations of physical locations, time, people, software agents, mobile devices, and meeting events. SOUPA[14] (Standard Ontology for Ubiquitous and Pervasive Applications) uses other standard domain ontologies, such as FOAF[4] (Friend of A Friend), OpenGIS, spatial relations in OpenCyc, ISO 8601 date and time formats[5], and DAML time ontology[15]. Clearly, these ontologies provide not only rich contextual representations, but also make use of the abilities of reasoning and sharing knowledge.

2.3 Context Reasoning

Design and implementation of context reasoning can vary depending on types of contextual information that are involved. Early context-aware systems[16,17,18] embedded the understanding of the contextual information into the program code of the system. Therefore, developed applications often had rigid implementations and were difficult to maintain.

Rule-based logical inference can help to develop flexible context-aware systems by separating high-level context reasoning from low-level system behavior. However, context modeling languages are used to represent contextual information and rule languages are used to enable context reasoning. Accordingly, in

---

[1] http://www.mindswap.org/2004/geo/geoStuff.shtml
[2] http://tools.ietf.org/html/rfc2445
[3] http://calendar.google.com
[4] http://xmlns.com/foaf/spec/
[5] http://www.w3.org/TR/NOTE-datetime

most cases, these two types of languages have different syntaxes and semantic representations; thus it is a challenge to effectively integrate these distinctive languages in the development of context-aware systems. A mechanism to convert between contextual modeling and reasoning languages is one of solutions for this challenge. Gandon and Sadeh[19,20] propose e-Wallet which implements ontologies as context repositories and uses a rule engine Jess[21] to invoke the corresponding access control rules. The e-Wallet using RDF[6] triples to represent contextual information and OWL[7] to define context ontology. Contextual information is loaded into the e-Wallet by using a set of XSLT[8] stylesheets to translate OWL input files into Jess assertions and rules.

Ontology models can represent contextual information and specify concepts, subconcepts, relations, properties, and facts in a smart space. Moreover, ontology reasoning can use these relations to infer the facts that are not explicitly stated in the knowledge base. Ranganathan et al.[22] assert that ontologies make it easier to develop programs for reasoning about context. Chen[23] proposes that the OWL language can provide a uniformed solution for context representation and reasoning, knowledge sharing, and meta-language definitions. Anagnostopoulos et al.[24] name the Description Logic as the most useful language for expressing and reasoning contextual knowledge. The OWL DL was designed to support the existing Description Logic business segment and to provide desirable computational properties for reasoning systems. Typical ontology-based context-aware application is Smart Meeting Room that uses OWL to define the SOUPA ontology and OWL DL to support context reasoning[3],Gu et al.[25] propose an OWL encoded context ontology CONON in Service Orientated Context Aware Middleware (SOCAM). CONON consists of two layers of ontologies, an upper ontology that focuses on capturing general concepts and a domain specific ontology. The Smart Meeting Room and SOCAM use an OWL DL reasoning engine to check the consistency of contextual information and provide further reasoning over low-level contexts to derive high-level contexts.

3   System Architecture

Fig. 1 shows our Context-aware System Architecture, which can continuously adapt to changing contexts and proactively provide services to the user. The top part of Fig. 1 depicts a smart space environment, which is equipped with devices and applications, such as a personal calendar, weather forecast, location tracking system, contact list, shopping list, as well as raw sensing data. The system can provide both contextual information and deliver context-aware services. The Context Collection Agents obtain raw sensing data from the context sources and convert the raw context into a semantic representation. Each Context Collection Agent will deliver the sensed contextual information to the *Context Management* component after receiving sensing data.

---

[6] http://www.w3.org/TR/rdf-concepts/
[7] http://www.w3.org/TR/owl-features/
[8] http://www.w3.org/TR/xslt

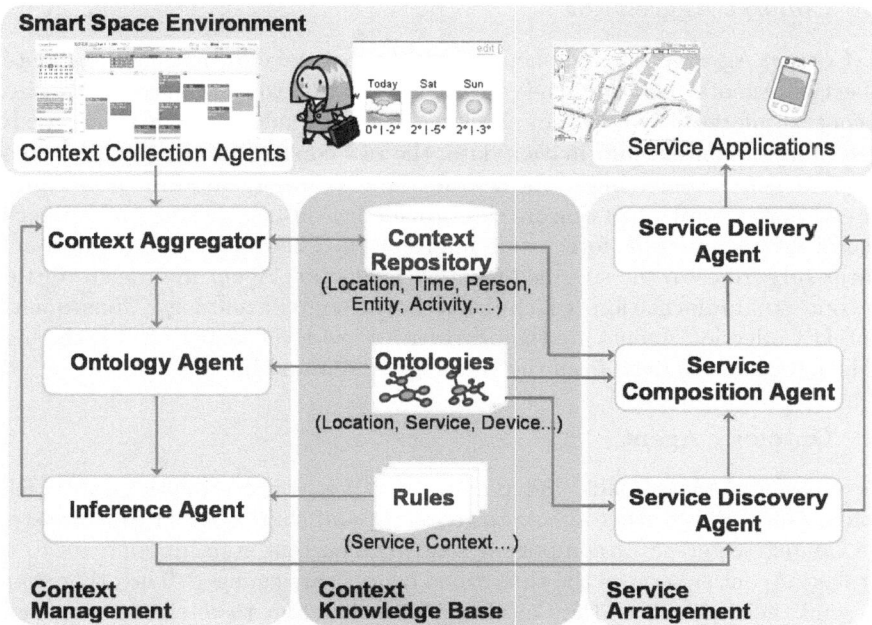

Fig. 1. A Multi-agent System Architecture

The lower part of Fig. 1 illustrates our context-aware system architecture, which consists of three components: *Context Management, Context Knowledge Base*, and *Service Arrangement*. The functions of *Context Management* component include monitoring the contextual information and managing the environmental resources. Contextual information, domain knowledge, and service profiles are stored in the *Context Knowledge Base*. The *Service Arrangement* component performs service discovery, composition, and execution. After assigning the specified services, the system will invoke Service Applications to provide services in the smart space environment.

4   Context-Aware Agents

Agents in Fig. 1 accomplish the functions of managing contextual information and delivering context-aware services. Functions of *Context Management* component include gathering context information from the surrounding environment, providing methods for querying and storing the contextual information, and providing the context in an ontology-based representation that facilities knowledge representation and inference. The *Service Arrangement* module checks whether any service operations match the request under the current situation. If no operation matches the request, it combines operations to match the request.

4.1   Context Aggregator

The Context Aggregator component collects contextual information from Context Collection Agents and stores the context information to the Context Repository for context inference, consistency checking, and knowledge sharing. There are two types of input context information data, the raw context and the high-level context. Raw context referring to the sensing data is directly obtained from context sources. For example, bed sensors can provide lay-on-bed sensing and a weather forecast API can provide forecasting information. The Context Aggregator component subscribes to the specified Context Collection Agent in order to retrieve the contextual information, which define in the context ontology. Consequently, Context Collection Agents are the providers of low-level contexts while the high-level contexts are derived from the Ontology Agent and Inference Agent.

4.2   Ontology Agent

The Ontology Agent loads and parses an OWL context ontology into RDF triples, which allows other agents to represent and share context in the system. The Context Aggregator component sends the current state of contexts to the Ontology Agent as soon as the subscribed context are changed. The other agents can send their queries to the Ontology Agent in order to retrieve the updated knowledge. According to the structures and relationships between contexts that define in the context ontology, the Ontology Agent performs the subsumption reasoning to deduce new contextual information. For example, it can deduce the superclasses of a specified class and decide whether one class consists of members of another, *e.g.*, a building may spatially contain a room.

4.3   Inference Agent

The Inference Agent adopts an OWL DL reasoning engine for supporting context reasoning and conflict detection. When the Inference Agent receives contextual information from the Ontology Agent, the reasoning engine will invoke rules and trigger actions that may deduce new high-level contexts and derive associated service requests. Combining the inferred contexts with the original context ontology, the Inference Agent can detect the context inconsistency. Either new high-level contexts or service requests can be derived from the Inference Agent and be delivered to the Context Aggregator or the Service Discovery Agent, respectively.

4.4   Service Discovery Agent

The Service Discovery Agent maintains the service ontology. An OWL-S[9] file defines the service ontology, which includes three essential types of knowledge about a service: the service profile, process model, and service grounding. The OWL-S service profile illustrates the preconditions required by the service and

---

[9] http://www.w3.org/Submission/OWL-S/

the expected effects that result from the execution of the service. A process model describes how services interact and functionalities are offered, which can be exploited to solve goals. The role of service grounding is to provide concrete details of message formats and protocols. According to the description in the service ontology, the Service Discovery Agent keeps the atomic services information. When the Service Discovery Agent receives a service request, it checks whether any single service satisfies the requirement under the current situation. If an atomic service can accomplish the request, the associated service grounding information will be delivered to the Service Delivery Agent.

4.5 Service Composition Agent

If a service request cannot be achieved by a single service, the Service Composition Agent will compose atomic services to fulfill the request. The service profile of a service ontology defines the service goals, preconditions, and effects. According to these semantic annotations, AI planning has been investigated for composing services. The state transition is defined by the operations, which consist of preconditions and effects. Initial states of the AI planner are combined with the current contexts and context ontology. The service request is the planning goal. Therefore, giving initial states, goals, and operations, the Service Composition Agent will derive a service execution plan, which is a sequence of operations that starts from initial states and accomplishes the given goal.

4.6 Service Delivery Agent

The service ontology defines the information for service grounding, which specifies the details of how an agent can access a service. According to the description of service grounding, the Service Delivery Agent invokes the specified Service Application with the required protocol and message contents.

5 Context Ontology Model

Context-aware applications need a unified context model that is flexible, extendible, and expressive to adapt to a variety of context features and dependency relations. The ontology models can fulfill these requirements; therefore, we deploy an ontology context model to represent contextual information in smart space environments. The ontology fills the need to share knowledge about locations, time, and activities so that context-aware applications can infer the environmental contexts and trigger services.

5.1 Context Repository

The Context Repository stores a set of consistent context, which including location, person, and activity information. Both raw and high-level context have a unique type identities and values. The associated value in the timestamp represents when the corresponding context arrived. The context ontology defines the classes of contexts and the relationships between the instances of context objects. A RDF-triple represents a context that contains a subject, a predicate, and an object. The subject is a resource named by a URI with an optional anchor identity. The predicate is a property of the resource, while the object is the value of the property for the resource. For example, the following triple represents "Peter is sleeping".

```
<http://...#Peter>
<http://...#participatesIn>
<http://...#sleeping>
```

Where `Peter` represents a subject, `participatesIn` is a predicate, and the activity `sleeping` is an object. We use subject and predicate as the compound key of the Context Repository. When a context has been updated, the associated timestamp will be changed accordingly.

5.2   Ontologies

An ontology is a data model that represents a domain and is used to reason about the objects in that domain and their relations. We define a context ontology (as depicted in Fig. 2) as a representation of common concepts about the smart space environment. Context information is collected from real-world classes (Person, Location, Sensor, Time, HomeEntity), and a conceptual class Activity. The class hierarchy represents an `is-a` relation; an arrow points from a subclass to a

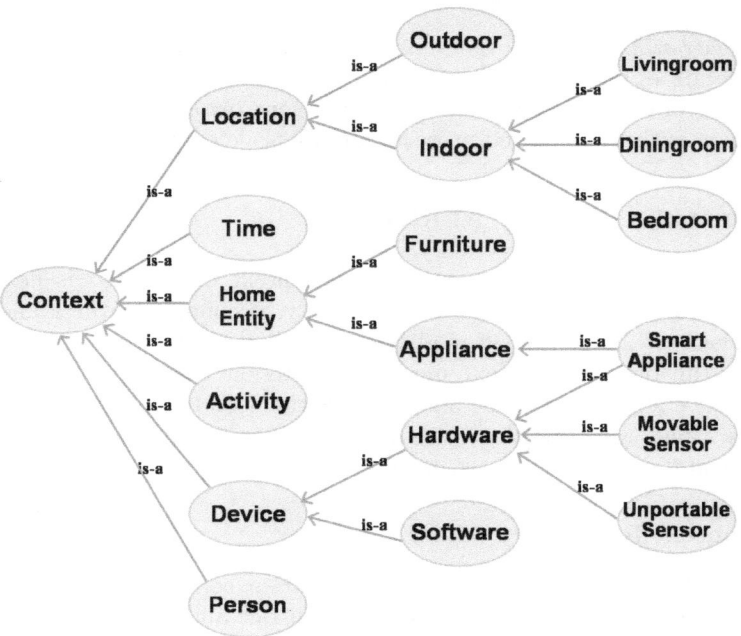

Fig. 2. A Context Ontology

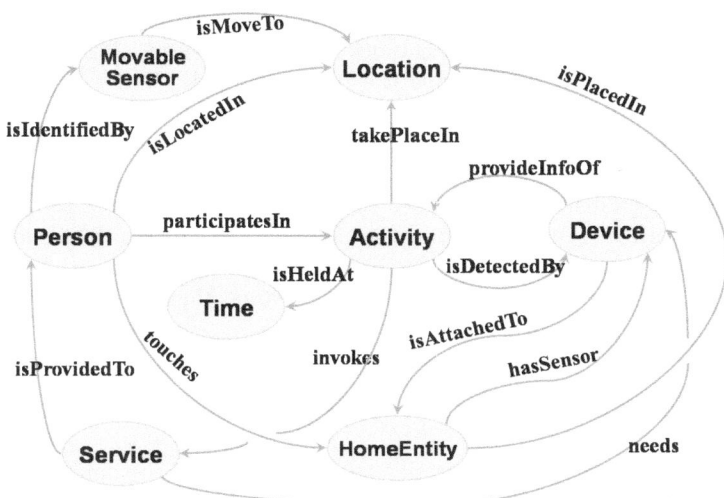

Fig. 3. Context Relationship

superclass. A class can have subclasses that represent the concepts more specific than itself. For example, we can divide the classes of all locations into indoor and outdoor locations, that is, Indoor Location and Outdoor Location are two disjoint classes and both of them belong to Location class. In addition, the subclass relation is transitive, therefore, the Livingroom is a subclass of Location class because Livingroom is a subclass of Indoor and Indoor is a subclass of Location.

The relationship between classes is illustrated in Fig. 3. The solid arrows describes the relation between subject resources and object resources. For example, isLocatedIn describes the relation between the instances of Person and Location while the instances of Person are subject resources and instances of Location are object resources.

A service ontology defined by OWL-S describes available services that are comprised of the service profile, service model, and service grounding.

5.3 Rules

Rules of a rule-based system are simply IF-THEN statements. Context rules can be triggered to infer high-level context. According to the description of Fig. 3, a rule for detecting the location of a user is as follows:

```
[Person_Location:
  (?person isIdentifiedBy ?tag)
  (?tag isMoveTo ?room)
->
  (?person isLocatedIn ?room )
]
```

Patterns before `->` are the conditions, matched by a specific rule, called the left hand side (LHS) of the rule. On the other hand, patterns after the `->` are the statements that may be fired, called the right hand side (RHS) of the rule. If all the LHS conditions are matched, the actions of RHS will be executed. The RHS statement can either infer a new high-level context or deliver a service request.

The rule `Person_Location` is an example that can deduce high-level context. The `?person` is an instance of class Person, `?tag` is an instance of MovableSensor, and `?room` is an instance of Room, the rule `Person_Location` declares that if any person `?person` is identified by a movable sensor `?tag` and this movable sensor is moved to a room `?room`, we can deduce that `?person` is located in `?room`.

6   Context Management and Reasoning Mechanism

In order to make our research easier to explain, we use a simple example to describe the detailed mechanisms of context management and reasoning.

In a smart space, we find a Smart Alarm Clock and a typical user whom we call Peter. The Smart Alarm Clock can check Peter's schedule and automatically set the wake-up alarm so as he will not miss his first appointment. If Peter does not wake up within a 5-minute period after the alarm is sounded, another alarm event is triggered, this time with increased volume. Alternatively, if Peter were to wake up earlier then the alarm time, the alarm would be disabled.

6.1   Context Reasoning

Before implementing the *Smart Alarm Clock*, we must collect Peter's schedule to ascertain appropriate alarm times and employ reasoning as to whether Peter is awake or not. The Google Calendar Data API[10] can support the information of Peter's calendar events. The position-aware sensors, bed pressure sensors, *etc.* can help to detect whether the user is on the bed or not. For example, RFID technologies can be used to recognize and identify Peter's activities. A wireless-based indoor location tracking system determines Peter's location with room-level precision.

Fig. 4 shows the instance relationships used in detecting whether Peter is currently sleeping or not. Words within ovals represent classes and boxes represent corresponding instances of these classes. For example, bed is an instance of the Furniture class. Dashed lines indicate the connection of a class and its instance. Each solid arrow reflects the direction of object property relationship and is directed from domain to range. In addition, an inverse property can be declared by reversing the direction of a line. For example, the inverse object property of `isAttachedTo` is `hasSensor`.

---

[10] `http://code.google.com/apis/calendar/`

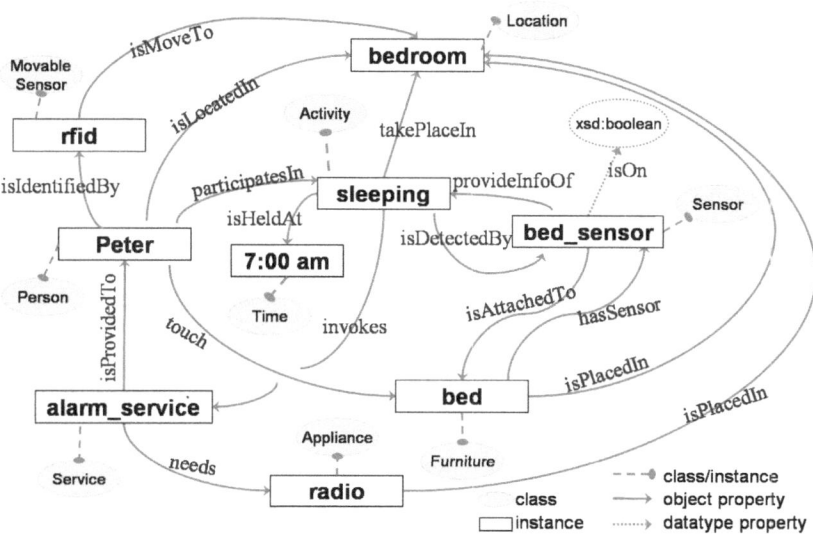

Fig. 4. A Context Snapshot

A boolean data type property isOn is associated with the Sensor class for detecting whether the value of instances are on or off. If someone is on the bed, the value of the bed sensor bed_sensor would be on, that is, the value of isOn would be **true**. Otherwise, when no one occupies the bed, the value of isOn would be **false**. When wake-up calls event has been triggered, a rule for detecting the value of the bed_sensor can be used to decide whether it is necessary to deliver the alarm_service or not.

For reasoning about high-level contexts, we apply rule-based reasoning with horn clauses into the ontology model. The rule **Person_activity** can deduce in which activity the user currently is involved. For example, in Fig. 4, when the time is up, given the location of Peter and the status of the bed sensor, the rule **Person_activity** will be triggered and can deduce whether Peter is sleeping or not. The rule **Invoke_service** can deduce what service should be delivered to the user. Given the instances of Fig. 4, the rule **Invoke_service** reflects "if Peter is sleeping, deliver smart alarm service".

```
[Person_activity:
  (?person touch ?entity)
  (?entity hasSensor ?sensor)
  (?sensor providesInfoOf ?activity)
->
  (?person participatesIn ?activity) ]
```

```
[Invoke_service:
  (?person participatesIn ?activity)
  (?activity invokes ?service)
->
  (?service isProvidedTo ?person)]
```

6.2 Context Management

Changes of environmental contexts are transient in the sense that context may appear and vanish at anytime. Algorithm 1 shows how the Context Aggregator manages the contextual information.

Algorithm 1. Maintaining Context Repository

1: **Input:** c is the new context
2: C: Context Repository
3: rdf_i: RDF-triple (s_i, p_i, o_i) of a context i
4: key_i: key of context i in Context Repository
5: **for all** $i \in C$ **do**
6:     **if** $isOutdated(i)$ **then**
7:        $delete(i)$
8:     **end if**
9: **end for**
10: **if** $\exists i \in C$ s.t. $key_i = key_c$ and $isOne2One(p_c)$ **then**
11:     $update(key_c, c)$
12: **else**
13:     $insert(key_c, c)$
14: **end if**

We use RDF-triple to represent a context while an associated compound key comprises the subject and object. When a new context arrives, the Context Aggregator uses the key of the new context to query the Context Repository. If a context exists in the Context Repository and the associated predicate represents a one-to-one relationship, the new context will replace the old one. Otherwise, the new context will be inserted into the Context Repository. Functions $update(key_c, c)$ and $insert(key_c, c)$ perform the context replacement and insertion, respectively. When a context is no longer applicable, it should be removed. function $delete(i)$ can remove the specified context from the Context Repository. We use a decay function to determine the existence of a given context. Different contexts are associated with different decay functions. This function can either be an objective function for predicating a specified activity or simply be a constant function. The function $isOutdated(i)$ applies the context decay function to decide whether the context exists or not.

6.3 Inconsistency Resolution

The Context Repository is dynamically updated to reflect the change of context. Therefore, we must ensure incorrect or outdated contexts do not exist in the

Context Repository. If a raw context is changed, some of the inferred high-level contexts may also be changed. For example, if Peter walks from the living room to the bedroom, the corresponding RDF-triple will be changed from `<Peter isLocatedIn living_room>` to `<Peter isLocatedIn bedroom>`. The Context Aggregator will update the location context of Peter because the property isLocatedIn is one-to-one relationship. If a predicate allows one-to-many relationship, the original context will be preserved.

It is a challenge that when a raw context is changed, we need to updated the associated high-level contexts. However, it is often difficult to find the corresponding high-level contexts using the context dependencies of inference rules. Updating a context may easily trigger an infinite context dependency checking loop and can lead to unpredictable situations. To address this issue, we categorize the data in the Context Repository to three types: core knowledge, raw-level contexts, and high-level contexts. The OWL ontologies define core knowledge that is static and persistent. The raw-level context is the raw sensor data that is delivered by the Context Collection Agents in Fig. 1. Using the core knowledge and raw-level contexts, rule-based reasoning can deduce high-level contexts. When a raw context has been removed, we discard the original set of high-level contexts and perform context reasoning. By design, the deduced high-level contexts are consistent with the current raw contexts and the Context Repository can maintain context consistency. This approach is while simple, but can efficiently resolve context inconsistency without recursively checking context dependencies.

7   Implementation

Our agent is deployed on JADE[11] (Java Agent DEvelopment Framework), which is a FIPA-compliant software framework for multi-agent systems, implemented in Java and comprised of several agents. Jena[12], a Java framework for building Semantic Web applications, is used for providing a programmatic environment for RDF, RDFS, and OWL. Moreover, we use an open-source OWL Description Logics (OWL DL) reasoner Pellet[13] developed by Mindswap Lab at University of Maryland, to infer high-level contexts and detect context conflicts.

8   Conclusion and Future Work

This research presents a context management mechanism in a smart space. We integrate context-aware technologies, semantic web, and logical reasoning to provide context-aware services. An ontology-based model supports a reasoning mechanism, which can deduce high-level contexts and detect context consistency.

We use a simple scenario to demonstrate the mechanism of context management. However, this simple case does not fully illustrate the power of context reasoning. Therefore, design of other scenarios that can better explain and evaluate our approach is one of our future directions.

---

[11] http://jade.tilab.com/

[12] http://jena.sourceforge.net/

[13] http://pellet.owldl.com/

References

1. Abowd, G.D., Atkeson, C.G., Bobick, A.F., Essa, I.A., MacIntyre, B., Mynatt, E.D., Starner, T.E.: Living laboratories: the future computing environments group at the georgia institute of technology. In: Proceedings of Conference on Human Factors in Computing Systems (CHI 2000): extended abstracts on Human factors in computing systems, pp. 215–216. ACM Press, New York (2000)
2. Intille, S.S.: Designing a home of the future. IEEE Pervasive Computing 1(2), 76–82 (2002)
3. Chen, H., Finin, T., Joshi, A., Kagal, L., Perich, F., Chakraborty, D.: Intelligent agents meet the semantic web in smart spaces. IEEE Internet Computing 8(6), 69–79 (2004)
4. Look, G., Shrobe, H.: A plan-based mission control center for autonomous vehicles. In: IUI 2004: Proceedings of the 9th international conference on Intelligent user interfaces, pp. 277–279. ACM Press, New York (2004)
5. Long, S., Aust, D., Abowd, G., Atkeson, C.: Cyberguide: prototyping context-aware mobile applications. In: Conference companion on Human factors in computing systems (CHI 1996), pp. 293–294. ACM Press, New York (1996)
6. Want, R., Hopper, A., Falcão, V., Gibbons, J.: The active badge location system. ACM Transactions on Information Systems (TOIS) 10(1), 91–102 (1992)
7. Want, R., Schilit, B.N., Adams, N.I., Gold, R., Petersen, K., Goldberg, D., Ellis, J.R., Weiser, M.: An overview of the PARCTAB ubiquitous computing experiment. Personal Communications 2(6), 28–43 (1995)
8. Harter, A., Hopper, A., Steggles, P., Ward, A., Webster, P.: The anatomy of a context-aware application. Wireless Networks 8(2-3), 187–197 (2002)
9. Dey, A.K.: Providing architectural support for building context-aware applications. PhD thesis, Georgia Institute of Technology, Director-Gregory D. Abowd (2000)
10. Ye, J., Coyle, L., Dobson, S., Nixon, P.: A unified semantics space model. In: Hightower, J., Schiele, B., Strang, T. (eds.) LoCA 2007. LNCS, vol. 4718, pp. 103–120. Springer, Heidelberg (2007)
11. Strang, T., Linnhoff-popien, C.: A context modeling survey. In: Workshop on Advanced Context Modelling, Reasoning and Management at The Sixth International Conference on Ubiquitous Computing (UbiComp 2004), Nottingham, England (2004)
12. Gruber, T.R.: A translation approach to portable ontology specifications. Knowledge Acquisition 5(2), 199–220 (1993); Special issue: Current issues in knowledge modeling
13. Chen, H., Finin, T., Joshi, A.: An ontology for context-aware pervasive computing environments. The Knowledge Engineering Review 18(3), 197–207 (2003)
14. Chen, H., Perich, F., Finin, T., Joshi, A.: SOUPA: Standard ontology for ubiquitous and pervasive applications. In: The First Annual International Conference on Mobile and Ubiquitous Systems: Networking and Services (MobiQuitous 2004), pp. 258–267 (August 2004)
15. Hobbs, J.R., Pan, F.: An ontology of time for the semantic web. ACM Transactions on Asian Language Information Processing (TALIP) 3(1), 66–85 (2004); Special Issue on Temporal Information Processing
16. Coen, M.H.: Building brains for rooms: designing distributed software agents. In: Proceedings of the Conference on Innovative Applications of Artificial Intelligence (IAAI 1997), pp. 971–977. AAAI Press, Menlo Park (1997)

17. Wu, H., Siegel, M., Ablay, S.: Sensor fusion for context understanding. In: Proceedings of IEEE Instrumentation and Measurement Technology Conference, Anchorage, AK, USA, May 21-23 (2002)
18. Capra, L., Emmerich, W., Mascolo, C.: CARISMA: Context-aware reflective mIddleware system for mobile applications. IEEE Transactions on Software Engineering 29(10), 929–945 (2003)
19. Gandon, F.L., Sadeh, N.M.: A semantic E-wallet to reconcile privacy and context awareness. In: Fensel, D., Sycara, K., Mylopoulos, J. (eds.) ISWC 2003. LNCS, vol. 2870, pp. 385–401. Springer, Heidelberg (2003)
20. Gandon, F.L., Sadeh, N.M.: Semantic web technologies to reconcile privacy and context awareness. Journal of Web Semantics 1(3), 241–260 (2004)
21. Friedman-Hill, E.: Jess in Action: Java Rule-Based Systems. Manning Publications, Greenwich (2003)
22. Ranganathan, A., Al-Muhtadi, J., Campbell, R.H.: Reasoning about uncertain contexts in pervasive computing environments. IEEE Pervasive Computing 3(2), 62–70 (2004)
23. Chen, H.: An Intelligent Broker Architecture for Pervasive Context-Aware Systems. PhD thesis, University of Maryland, Baltimore County (2004)
24. Anagnostopoulos, C.B., Tsounis, A., Hadjiefthymiades, S.: Context awareness in mobile computing environments. Wireless Personal Communications: An International Journal 42(3), 445–464 (2007)
25. Gu, T., Wang, X.H., Pung, H.K., Zhang, D.Q.: An ontology-based context model in intelligent environments. In: Proceedings of Communication Networks and Distributed Systems Modeling and Simulation Conference, pp. 270–275 (2004)

Using THOMAS for Service Oriented Open MAS

V. Julian, M. Rebollo, E. Argente, V. Botti, C. Carrascosa, and A. Giret

Dept. Sistemas Informáticos y Computación
Universidad Politécnica de Valencia
Camino de Vera s/n
46022 - Valencia, Spain
{vinglada,mrebollo,eargente,vbotti,carrasco,agiret}@dsic.upv.es

Abstract. Recent technological advances in open systems have imposed new needs on multi-agent systems. Nowadays, open systems require open autonomous scenarios in which heterogeneous entities (agents or services) interact to fulfill the system goals. This impose the need for open architectures and computational models for large-scale open multi-agent systems based on service-oriented approaches. THOMAS is a new architecture specifically addressed for the design of virtual organizations for open systems. In this paper we present a case study that exemplifies the usage of THOMAS for implementing a management system of a travel agency.

1 Introduction

Over recent years, several works have focused on solving the problem of integrating the multi-agent system and the service-oriented computing paradigms in order to model autonomous and heterogeneous computational entities in dynamic, open environments. Two approaches can be identified: (i) direct integration of web services and agents using message exchange [1,2], and (ii) considering agents as matchmakers for service discovering and composition [5,6,7].

A key problem for open MAS development is the existence of real agent platforms that support organizational concepts. Many agent platforms and agent architectures have been proposed, some of them focused on organizational concepts [8], but the majority are lacking in the management of virtual organizations for dynamic, open and large-scale environments. Designers must implement nearly all of the organizational features by themselves, namely organization representation and descriptions, control mechanisms, AMS and DF extensions, monitoring, organization modeling support and organizational API.

In this paper, a new open multi-agent system architecture is defined, named THOMAS (Me*TH*ods, Techniques and Tools for *O*pen *M*ulti-*A*gent *S*ystems), consisting of a related set of modules that are suitable for the development of systems applied in open environments in which heterogeneous entities (agents and services) interact. The proposed solution tries to communicate agents and web services in an independent way, thus raising a total integration of both

R. Kowalczyk et al. (Eds.): SOCASE 2009, LNCS 5907, pp. 56–70, 2009.

multi-agent and service-oriented technologies. Therefore, agents can offer and invoke services to other agents or entities in a transparent way, as well as external entities can interact with THOMAS agents through the use of the offered services. Thus, THOMAS not only deals with the management of virtual organizations, such as in S-Moise+ [9] and Ameli (EI platform) [10], but it also includes a prototype platform fully addressed for open systems which has all organizational features in mind, trying to obtain a framework wholly independent of any internal agent platform.

This paper is structured as follows: Section 2 presents the THOMAS architecture model. The description of the services offered by the THOMAS main components are described in Sections 2.1 and 2.2. Section 3 shows a complete example of a travel agency case-study in which the overall functioning and the features of the THOMAS architecture can be observed. Finally, the conclusions are presented.

2   Thomas Architecture

THOMAS architecture basically consists of a set of modular services. Though THOMAS initially feeds on the FIPA architecture, it expands its capabilities to deal with organizations, and to boost up its service abilities. In this way, a new module in charge of managing organizations has been introduced into the architecture, along with a redefinition of the FIPA *Directory Facilitator* that is able to deal with services in a more elaborated way, following *Service Oriented Architectures* guidelines. The main components of THOMAS (Figure 2) are the following: *Service Facilitator* (SF), which offers simple and complex services to the active agents and organizations; *Organization Management System* (OMS), responsible of the management of the organizations and their entities; and *Platform Kernel* (PK) which maintains basic management services for an agent platform and represents any FIPA compliant platform. In the following two sections the SF and the OMS components are described.

2.1   Service Facilitator

The *Service Facilitator* (SF) is a mechanism and support by which organizations and agents can offer and discover services. The SF provides a place in which the autonomous entities can register service descriptions as directory entries.

The SF acts as a gateway to access the THOMAS platform. It manages this access transparently, by means of security techniques and access rights management. The SF can find services searching for a given service profile or searching by the goals that can be fulfilled when executing the service. This is done using the matchmaking and service composition mechanisms that are provided by the SF. It also acts as a yellow pages manager and in this way it can find which entities provide a given service.

A service can be supplied by more than one provider in the system. Thus, a service has an associated list of providers, who can offer exact copies of the service, i.e. they share a common implementation of the service; or they may only

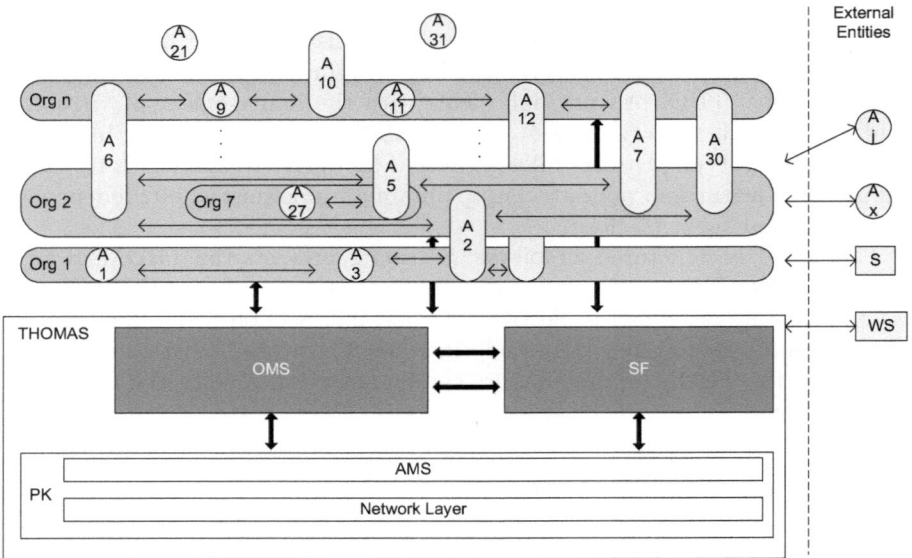

Fig. 1. THOMAS Architecture

share the service interface and each provider may implement it in a different way. This is easily achieved in THOMAS since the service profile is separated from the service process.

The SF supplies a set of standard services to manage the services provided by organizations or individual agents. These services can also be used by the rest of THOMAS components to advertise their own services. SF services are classified in: *Registration* (for adding, modifying and removing services from the SF directory); *Affordability* (for managing the association between providers and their services) and *Discovery* (for searching and composing services as an answer to user requirements).

2.2   Organization Management System

The *Organization Management System (OMS)* is in charge of the organization life-cycle management, including specification and administration of both the structural components of the organization (roles, units and norms) and its execution components (participant agents and roles they play).

Organizations are structured by means of *organizational units* (OUs), which represent groups of entities (agents or other units), that are related in order to pursue a common goal. OUs can also be seen as virtual meeting points because agents can dynamically enter and leave them by means of adopting (or leaving) roles inside. Roles represent all required functionality needed in order to achieve the unit goal. They might also have associated norms for controlling role actions. Agents can dynamically adopt roles inside units, so a control for role adoption is needed. Finally, services represent some functionality that agents offer to other entities, independently of the concrete agent that makes use of them.

The OMS keeps record on which are the Organizational Units of the system, the roles defined in each unit and their attributes, the entities participating inside each OU and the roles that they enact through time. Moreover, the OMS also stores which are the norms defined in the system.

The OMS offers a set of services for organization life-cycle management, classified in: *Structural* (for modifying the structural/normative organization specification); *Informative* (for informing of the current state of the organization); and *Dynamic* (for managing dynamic entry/exit of agents and role adoption). Thus, it includes services for creating new organizations, admitting new members within those organizations and member resigning. By means of the publication of the *structural services*, the OMS allows the modification of some aspects related to the organization structure, functionality or normativity at execution time.

3   Using THOMAS

In order to illustrate the performance of THOMAS platform with greater detail, a case-study example for making flight and hotel arrangements is used. This is a well known example that has been modelled by means of electronic institutions in previous works [11,12].

The *Travel Agency* example is an application that facilitates the interconnection between clients (individuals, companies, travel agencies) and providers (hotel chains, airlines); delimiting services that each one can offer and/or request. The system controls which services must be provided by each agent. Internal functionality of these services is responsibility of provider agents. However, the system imposes some restrictions about service profiles, service requesting orders and service results.

Following, a description of the structure elements of the Travel Agency organization is explained. Later, in section 3.2, a dynamical usage of the organization is detailed, providing different execution scenarios.

3.1   Case-Study Organization Structure

This case study is modelled as an organization (*TravelAgency*) inside which there are two groups of agents (*HotelUnit* and *FlightUnit*). Each unit is dedicated to hotels or flights, respectively.

Three kind of roles can interact in the Travel Agency example: customer, provider and payee roles. The *Customer* role requests services of the system. More specifically, it can request hotel or flight search services, booking services for hotel rooms or flight seats, and payment services. The *Customer* role is specialized into two subroles according to each type of product (*HotelCustomer* and *FlightCustomer*). The *Provider* role is in charge of performing services. A *provider* agent offers hotel or flight search services; and, in some cases, booking hotel rooms or flight seats. The *provider* role is also specialized into *HotelProvider* and *FlightProvider*. Finally, the *Payee* role gives the advanced payment service. It represents

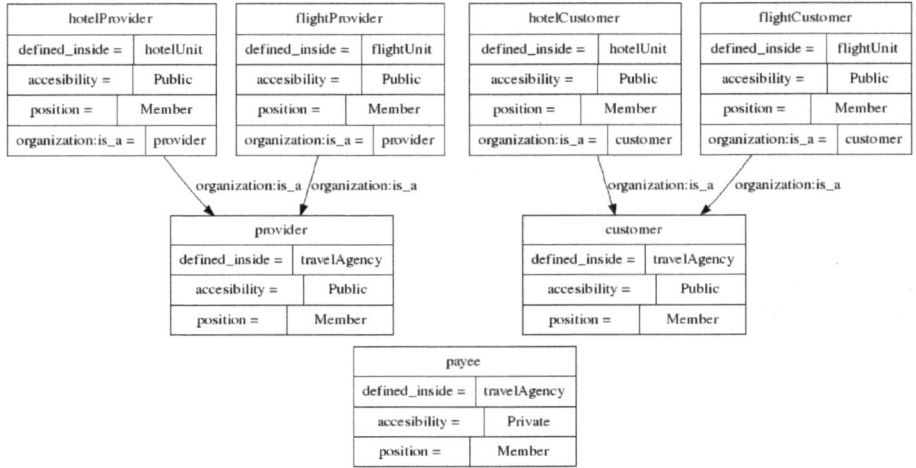

Fig. 2. Travel Agency role instances

a bank institution through which reservation payment is carried out. It is an internal role, so agents are not able to acquire this *Payee* role. Figure 2 shows the definition of *TravelAgency* roles and relationships between them.

The OMS component internally stores the list of defined units (*UnitList* table 1) and the *RoleList* that contains the list of roles (table 2).

Table 1. Content of *UnitList* for structural specification of the system

UnitList			
UnitName	**ParentUnit**	**Goal**	**Type**
Virtual(world)	—	—	Flat
TravelAgency	Virtual	TravelReserve	Congregation
HotelUnit	TravelAgency	ReserveHotel	Flat
FlightUnit	TravelAgency	ReserveFlight	Flat

The *TravelAgency* organization offers three services: *SearchTravel*, *Booking* and *Payment* service. *TravelAgency* services are specialized, inside each unit, according to each type of product (hotels and flights). Consequently, services registered inside *HotelUnit* are *SearchHotel* and *ReserveHotel*. Similarly, specialized services over flight domain are defined inside *FlightUnit*.

In addition to service identification, designers should provide a description of each service offered inside the *TravelAgency*. Service specification consists of an OWL-S service description. Service profile has been extended with client and provider roles and also with an identifier of the unit inside which this service is provided. Extended service profiles of *TravelAgency* services are contained within table 3. Tables 4 and 5 contain a brief description of services offered inside *HotelUnit* and *FlightUnit*, respectively.

Table 2. *RoleList* content

RoleList				
RoleName	**DefinedIn**	**Accesibility**	**Position**	**IsA**
Customer	TravelAgency	Public	Member	—
Provider	TravelAgency	Public	Member	—
Payee	TravelAgency	Private	Member	—
HotelCustomer	HotelUnit	Public	Member	Customer
HotelProvider	HotelUnit	Public	Member	Provider
FlightCustomer	FlightUnit	Public	Member	Customer
FlightProvider	FlightUnit	Public	Member	Provider

Table 3. TravelAgency Service Profiles

Service		Description		UnitID
SearchTravel		Search for travel information		TravelAgency
ClientRole	**ProviderRole**	**Inputs**	**Outputs**	
Customer	Provider	city:string country:string	[city ok] company:string location:string price:float	[not in city] error
Service		**Description**		**UnitID**
Reserve		Make a reservation		TravelAgency
ClientRole	**ProviderRole**	**Inputs**	**Outputs**	
Customer	Provider	company:string location:string date:time numRsv:integer	[available] RsvTicket:Reserve price:float [advanced pay-ment] IBAN:integer	[not available] "completed"
Service		**Description**		**UnitID**
Payment		Advanced payment of a reservation		TravelAgency
ClientRole	**ProviderRole**	**Inputs**	**Outputs**	
Customer	Payee	RsvTicket:Reserve amount:float IBAN:string bankcard:string	[ok] Invoice:string	[error] bankError

The SF component stores information about services defined in THOMAS platform. Table 6 shows the content of *ServiceList* after service registering process. Initially, the *TravelAgency* has none agent registered as service *Provider*, thus no service can be offered. Consequently, *ProviderList* field of each registered service is empty.

3.2 System Dynamics

In this section, the use of THOMAS meta-services in *TravelAgency* example is detailed. System dynamics are shown through the specification of different scenarios: (i) an agent joins THOMAS platform; (ii) a *Provider* is registered; (iii) the provider is registered as a *HotelProvider*; (iv) definition of a new service implementation; (v) addition of new service providers; (vi) client registering; (vii) service requesting; (viii) expulsion of malicious agents; (ix) *Provider* deregistering; and (x) unit creation.

Table 4. Hotel Service Profiles

Service		Description		UnitID
SearchHotel		Search for inf. about hotels in a city		HotelUnit
ClientRole	ProviderRole	Inputs	Outputs	
HotelCustomer	HotelProvider	city:string country:string category:integer	[city ok] HotelChain:string HotelName:string RoomRate:float Address:string	[not in city] error
Service		Description		UnitID
ReserveHotel		Make a hotel reservation		HotelUnit
ClientRole	ProviderRole	Inputs	Outputs	
HotelCustomer	HotelProvider	company:string location:string date:time numRsv:integer nights:integer	[available] RsvTicket:Reserve price:float [advanced payment] IBAN:string	[not available] "completed" OptDate:date

Table 5. Flight Service Profiles

Service		Description		UnitID
SearchFlight		Search for information about flights		HotelFlight
ClientRole	ProviderRole	Inputs	Outputs	
FlightCustomer	FlightProvider	cityFrom:string countryFrom:string cityTo:string countryTo:string date:time	[cityFrom, cityTo ok] FlightCompany:string FlightType:string price:float HourDept:time HourArr:time NumConnect:integer	[not in city] error
Service		Description		UnitID
ReserveFlight		Make a flight reservation		HotelFlight
ClientRole	ProviderRole	Inputs	Outputs	
FlightCustomer	FlightProvider	company:string location:string numRsv:integer date:time cityTo:string HourDept:time HourArr:time	[available] RsvTicket:Reserve price:float [advanced payment] IBAN:string	[not available] "completed"

Table 6. *ServiceList* content

ServiceFacilitator		
ServiceID	Profile	ProvidersList
SearchTravel	SearchTravelPF	
Reserve	ReservePF	
SearchHotel	SearchHotelPF	
ReserveHotel	ReserveHotelPF	
SearchFlight	SearchFlightPF	
ReserveFlight	ReserveFlightPF	
Payment	PaymentPF	

Agent Registering. This scenario details the sequence of services that an agent should request in order to join the THOMAS platform. More specifically, CH1 is an agent that represents a hotel chain. Its functionality allows it to offer an information service about hotels belonging to its company. Agents join THOMAS platform by requesting OMS for being a *Member* of the *Virtual* organization (using *AcquireRole* service):

AcquireRole("Virtual", "Member")

OMS checks all restrictions (existence of unit and role identifiers, compatibility of role, etc.) and registers CH1 agent as a new member of THOMAS platform. OMS makes use of *RegisterAgentRole* service for adding a new item *(EntityID, RoleID, UnitID)* to *EntityPlayList* (table 7).

Table 7. *EntityPlayList* content after CH1 agent is registered

EntityPlayList		
Entity	Unit	Role
PayAgent	TravelAgency	Payee
CH1	Virtual	Member

Provider Registering. In this scenario, the process for registering a new *Provider* is illustrated. After CH1 has been registered as a member of THOMAS platform; it should ask the SF which are those defined services that have a profile nearby to its own "information search service. This request is carried out via the SF *SearchService* as follows:

SearchService(SearchServiceProfile)

where *SearchServiceProfile* corresponds to the profile of the hotel search service implemented by CH1.

The SF returns service identifiers that satisfy these search requirements together with a *ranking value* for each service. *Ranking value* indicates the degree of suitability between a service and a specified service purpose. The returned list by SF is:

$$(<\text{SearchTravel}, 0.75>)$$

Then CH1 executes *InformService* in order to obtain detailed information about the *SearchTravel* service:

InformService("SearchTravel")

Service outputs are "service goal" and "profile". The SearchTravel profile specifies that service providers have to play a *Provider* role inside *TravelAgency*. Thus, *AcquireRole* service allows CH1 to request role adoption:

AcquireRole("Provider", " TravelAgency ")

AcquireRole service is carried out successfully, because *TravelAgency* is accessible from *Virtual* organization, thus CH1 is registered as a *Provider*.

HotelProvider Registering. Once provider registering has been detailed, the register of a hotel provider is illustrated. CH1 is able to provide a search service in the hotel domain. Therefore, it asks to SF whether exists a service with a closer profile, requesting *SearchService* to SF as before. Answer provided by SF is:

(<SearchTravel, 0.75>, <SearchHotel, 1>)

In this second case, SF returns both *SearchTravel* and *SearchHotel* since these two services are visible from *TravelAgency*. As indicated in the service result, *SearchHotel* service is more appropriated for CH1 functionality. Therefore, CH1 requests to SF information about this service via *InformService*. *SearchHotel* profile specifies that service providers must play *HotelProviders* inside *HotelUnit*. CH1 requests OMS to adopt *HotelProvider* role. *AcquireRole* service is carried out successfully, so CH1 agent is registered as a *HotelProvider*.

AcquireRole("HotelProvider", "HotelUnit")

Register a New Service Implementation. In this example, the sequence of actions that allows an agent to register its own implementation of a service is detailed. Supposing that CH1 has assumed *HotelProvider* role and it is interested on providing its own implementation of *SearchHotel* service. Therefore it registers itself as service provider in SF employing *RegisterProcess* service:

RegisterProcess("SearchHotel", "SearchHotelProcess", "SearchHotelGrounging", "CH1")

where *SearchHotelProcess* and *SearchHotelGrounding* correspond to service process and grounding. Figure 3 contains the specification of *SearchHotel* service process. Figure 4 shows *SearchHotel* grounding; this type of grounding specifies how a service can be requested by means of sending ACL messages.

searchHotelProcess		
process:hasInput =		?category
		?country
		?city
process:hasOutput =		?hotelCompany
		?hotel
process:hasClient =		hotelCustomer
		process:TheClient
process:performedBy =		hotelProvider
		ch1.thomas
		process:TheServer
service:describes =	searchHotelService	

Fig. 3. Example of *SearchHotel* process

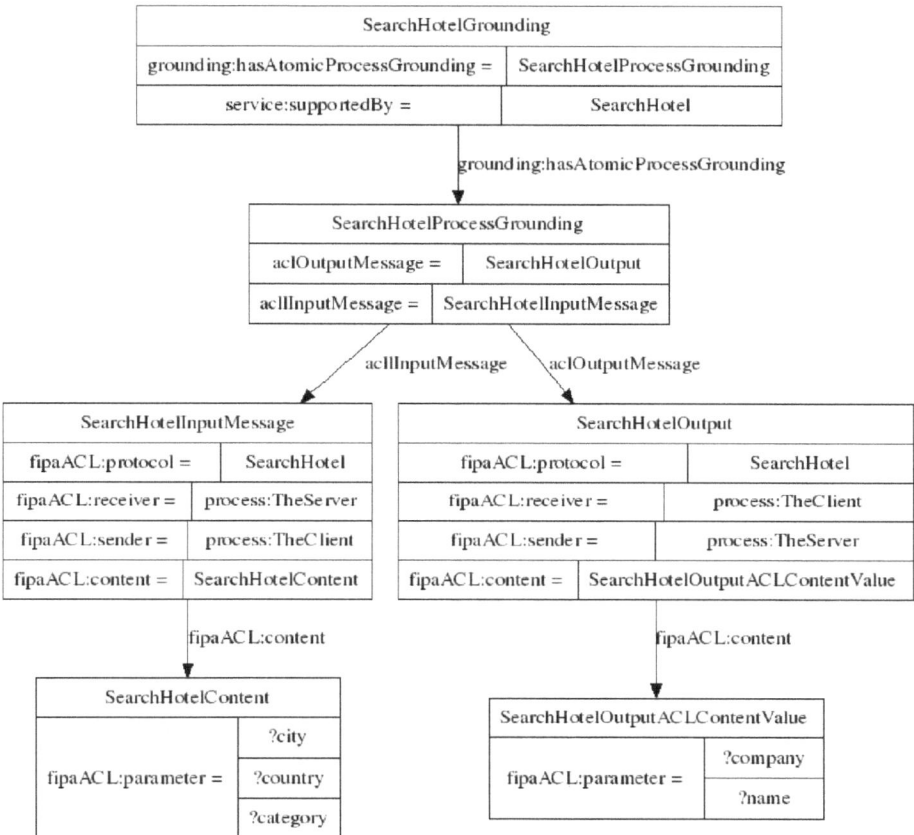

Fig. 4. Example of *SearchHotel* grounding

Addition of New Service Providers. This section exemplifies how another *HotelProvider* (CH2) is added to a service provider list. CH2 offers a service of *SearchHotel*, but this search is restricted to luxury hotels. Since *LuxurySearchHotel* service is performed in a similar way to *SearchHotel* service, CH2 decides to follow the registered implementation. Therefore, it uses *AddProvider* service in order to request its inclusion as provider of *SearchHotel* service:

$$AddProvider(\text{“SearchHotelImp”, “CH2”})$$

where *SearchHotelImp* is the identifier of the registered implementation of *SearchHotel*.

Client Registering. The following scenario shows the set of service calls for registering new agents as service clients inside the *TravelAgency*. The new client agent (C1) requests *SearchService* to SF for finding services of its interest:

$$SearchService(\text{“Search information about hotels”})$$

The result of this service is:

$$(<\text{SearchTravel, 0.25}>)$$

C1 employs *InformService* in order to know inside which unit *SearchTravel* service is provided. As indicated in the service profile, C1 must acquire *Customer* role for demanding this service. Once C1 plays this customer role, it employs *SearchProvider* service in order to know who are service providers and how this service can be requested.

SearchProvider("SearchTravel")

An empty list is provided as result, because any agent has been registered as service provider yet. Given that C1 can not request any service inside *TravelAgency*, it requests *SearchService* again to SF. Answer provided by SF is:

$$(<\text{SearchTravel, 0.25}>, <\text{SearchHotel, 1}>)$$

SF returns *SearchTravel* and *SearchHotel* services because both services are accessible from *TravelAgency* organization. C1 demands the profile of *SearchHotel* service (using *InformService*), since this service is more appropriated to its needs. In consideration of *SearchHotel* profile, C1 requests the adoption of *HotelCustomer* role inside *HotelUnit*

AcquireRole("HotelCustomer", "HotelUnit")

Service Requesting. This scenario shows how client agents make demands for services. Once C1 assumes client role for *SearchHotel* service, it is allowed to demand services to providers. Assuming that C1 wants to make an information search about hotels, it should use *SearchProvider* service for getting implementations of the service together with provider identifiers.

SearchProvider(" SearchHotel")

As it is shown in table 8, one implementation of *SearchHotel* has been previously registered. Therefore, answer to *SearchProvider* service provided by SF is:

$$<\text{SearchHotelGrounding, SearchHotelProcess, (CH1,CH2)}>,$$

After comparing providers of *SearchHotel* service, C1 chooses to make a service petition to CH1 agent. According to service process, ACL message sent by C1 for requesting *SearchHotel* service to CH1 agent is contained in table 9. In table 10 the service result sent by CH1 is shown.

Agent Expulsion. In this scenario, the expulsion of a malicious agent is carried out. *Payee* agent detects that different client agents (C1 and C2) have been paying their bookings with the same credit card number. It consults to its banking institution and it determines that C2 has been employing a credit card that does not belong to it. C2 is punished for its fraudulent behaviour and is expulsed from *TravelAgency*. *Payee* requests to OMS the expulsion of C2 employing *Expulse* service as follows:

Expulse("C2", "TravelAgency", "Customer")

Table 8. Content of ServiceList

ServiceFacilitator		
ServiceID	Profile	Providers
SearchHotel	SearchHotelPF	SearchHotelImp(CH1,CH2)
ReserveHotel	ReserveHotelPF	
SearchFlight	SearchFlightPF	
ReserveFlight	ReserveFlightPF	
SearchTravel	SearchTravelPF	
Reserve	ReservePF	
Payee	PaymentPR	PaymentImp(PayAgent)

Table 9. ACL message sent by C1 for requesting *SearchHotel* service

```
request
(
   :sender    c1.thomas
   :receiver  ch1.thomas
   :content
   (
               ...
               <city rdf:datatype="string">Valencia</city>
               <country rdf:datatype="string">Spain</country>
               <category rdf:datatype="integer">5</category>
               ...
   )
   :in-reply-to
   :language
   :ontology
   :protocol  SearchHotel
)
```

Table 10. ACL message sent by CH1 that contains *SearchHotel* result

```
inform
(
   :sender    ch1.thomas
   :receiver  c1.thomas
   :content
   (
               ...
               <name rdf:datatype="string">Hotel</name>
               <holtelCompany rdf:datatype="string">Chain</holtelCompany>
               ...
   )
   :in-reply-to
   :language
   :ontology
   :protocol  SearchHotel
)
```

Provider Deregistering. Following, the process of service provider deregistering is described. CH1 loses connection with its internal database. As a result, it is not able to provide services temporally. Therefore CH1 deregisters itself as provider of *SearchHotel* and *ReserveHotel* services, through requesting *Remove-Provider* to SF as follows:

$$RemoveProvider(\text{"SearchHotel", "CH1"})$$
$$RemoveProvider(\text{"ReserveHotel", "CH1"})$$

CH1 is deleted from both service provider lists. Nevertheless, as it continues playing *ProviderRole*, CH1 will be able to register itself as service provider if it recovers its functionality.

Unit Creation. This last scenario illustrates the creation of new units inside *TravelAgency*. Agent CR1 represents a tourism company that is specialized in cruisers. It is interested in providing information and booking services about cruises. Since offered services inside *HotelUnit* and *FlightUnit* are specialized over hotel and flight domains, CR1 decides to create a new unit (*CruiseUnit*) inside *TravelAgency*. This new unit will be focused on cruiser companies.

RegisterUnit(**"CruiseUnit", "Provider", "TravelAgency"**)

After OMS informs CR1 about the successful creation of the new unit, CR1 defines cruise specific roles and services:

RegisterRole(**"CuiseProvider", "CruiseUnit", "Public", "External", "Member"**)
RegisterRole(**"CruiseCustomer", "CruiseUnit", "Public", "External", "Member"**)
RegisterProfile(**"SearchCruise", "Search for information about cruisers", "SearchCruiseProfile"**)
RegisterProfile(**"ReserveCruise", "Make a cruiser booking", "ReserveCruiseProfile"**)

Once cruise unit, roles and services are created, cruiser agents can assume the *CruiserProvider* role and start to offer services to client agents.

4   Conclusions

An important aspect for the development of true open multi-agent systems is to provide developers with methods, tools and appropiated architectures which support all the requirements for this kind of systems. This paper has deepened into this problem trying to propose an abstract architecture for the development of virtual organizations. Moreover, the proposal tries to raise a total integration of two promising technologies, that is, multi-agent systems and service-oriented computing as the foundation of such virtual organizations. In THOMAS architecture, agents can offer and invoke services in a transparent way to other agents, virtual organizations or entities, as well as external entities can interact with agents through the use of the offered services.

A case-study example has been employed as illustration not only of the usage of THOMAS components and services, but also of the dynamics of the applications to be developed with such architecture. In this way, examples of THOMAS service calls have been shown through several scenarios, along with the evolution of the different dynamic virtual organizations happening in this kind of applications.

This architecture is the first step in order to obtain true deployed virtual organizations. Currently, a first version of a framework (v0.1) based on this proposal has being developed[1] and it is being applied in the development of different scenarios as tourism, leisure activity management on a mall and health emergencies.

Acknowledgments. The authors would like to thank the financial support received from the Spanish government and FEDER funds under CICYT TIN2006-14630-C03-01 and TIN2005-03395 projects is gratefully acknowledged. This project is also partially supported by CONSOLIDER-INGENIO 2010 under grant CSD2007-00022.

References

1. Greenwood, D., Lyell, M., Mallya, A., Suguri, H.: The IEEE fipa approach to integrating software agents and web services. In: AAMAS 2007: Proceedings of the 6th international joint conference on Autonomous agents and multiagent systems, pp. 1–7. ACM, New York (2007)
2. Shafiq, M.O., Ali, A., Ahmad, H.F., Suguri, H.: Agentweb gateway - a middleware for dynamic integration of multi agent system and web services framework. In: 14th IEEE International Workshops on Enabling Technologies (WETICE 2005), Linköping, Sweden, June 13-15 (2005), pp. 267–270. IEEE Computer Society, Los Alamitos (2005)
3. Varga, L.Z., Hajnal, Á.: Engineering web service invocations from agent systems. In: Mařík, V., Müller, J.P., Pěchouček, M. (eds.) CEEMAS 2003. LNCS, vol. 2691, pp. 626–635. Springer, Heidelberg (2003)
4. Nguyen, T., Kowalczyk, R.: Ws2jade: Integrating web service with jade agents. Technical Report SUTICT-TR2005.03, Centre for Intelligent Agents and Multi-Agent Systems, Swinburne University of Technology (2005)
5. Sensoy, M., Pembe, C., Zirtiloglu, H., Yolum, P., Bener, A.: Experience-based service provider selection in agent-mediated e-comerce. Engineering Applications of Artificial Intelligence 3, 325–335 (2007)
6. Caceres, C., Fernandez, A., Ossowski, S., Vasirani, M.: Role-based service description and discovery. In: International Joint Conference on Autonomous Agents and Multi-Agent Systems (2006)
7. Sycara, K., Paolucci, M., Soudry, J., Srinivasan, N.: Dynamic discovery and coordination of agent-based semantic web services. IEEE Internet Computing 8-3, 66–73 (2004)
8. Argente, E., Giret, A., Valero, S., Julian, V., Botti, V.: Survey of MAS Methods and Platforms focusing on organizational concepts. In: Vitria, J., Radeva, P., Aguilo, I. (eds.) Recent Advances in Artificial Intelligence Research and Development. Frontiers in Artificial Intelligence and Applications, pp. 309–316 (2004)
9. Hubner, J., Sichman, J., Boissier, O.: S-Moise+: A middleware for developing organised multi-agent systems. In: Boissier, O., Padget, J., Dignum, V., Lindemann, G., Matson, E., Ossowski, S., Sichman, J.S., Vázquez-Salceda, J. (eds.) ANIREM 2005 and OOOP 2005. LNCS, vol. 3913, pp. 64–78. Springer, Heidelberg (2006)

---

[1] More details can be found at http://www.dsic.upv.es/users/ia/sma/tools/Thomas

10. Esteva, M., Rodriguez-Aguilar, J., Sierra, C., Arcos, J., Garcia, P.: Lecture Notes in Artificial Intelligence 1991. In: On the Formal Specification of Electronic Institutions, pp. 126–147. Springer, Heidelberg (2001)
11. Dignum, F., Dignum, V., Thangarajah, J., Padgham, L., Winikoff, M.: Open Agent Systems? In: Luck, M., Padgham, L. (eds.) Agent-Oriented Software Engineering VIII. LNCS, vol. 4951, pp. 73–87. Springer, Heidelberg (2008)
12. Sierra, C., Thangarajah, J., Padgham, L., Winikoff, M.: Designing institutional multi-agent systems. In: Padgham, L., Zambonelli, F. (eds.) AOSE VII / AOSE 2006. LNCS, vol. 4405, pp. 84–103. Springer, Heidelberg (2007)

Agent-Based Support for Context-Aware Provisioning of IMS-Enabled Ubiquitous Services

Ana Petric[1], Krunoslav Trzec[2], Kresimir Jurasovic[1], Vedran Podobnik[1], Gordan Jezic[1], Mario Kusek[1], and Igor Ljubi[1]

[1] University of Zagreb
Faculty of Electrical Engineering and Computing
Unska 3, HR-10000, Zagreb, Croatia
{ana.petric,kresimir.jurasovic,vedran.podobnik,gordan.jezic,
mario.kusek,igor.ljubi}@fer.hr
[2] Ericsson Nikola Tesla
Krapinska 45, HR-10000, Zagreb, Croatia
krunoslav.trzec@ericsson.com

Abstract. Multimedia applications executed on mobile devices allow users to be present and communicate with other users, anywhere and anytime, through wide area cellular networks, wireless local area networks (WLAN), or fixed networks. In order to enable ubiquitous personalized services, communication systems should allow users to specify their context, or their mobile devices and network infrastructures should automatically sense their context and offer enhanced service provisioning solutions. In this paper, we propose an agent-based solution that supports context-aware provisioning of IP multimedia subsystem (IMS)-enabled ubiquitous services. Using agent technology, multimedia communication, controlled by SIP and enriched with context-related events delivered by SIP's event mechanism, can be seamlessly provisioned, taking into account not only mobility issues and context-awareness, but also the semantics of exchanged events. This paper proposes a multi-agent system that will enable users to consume context-aware ubiquitous services in a seamless (i.e., automated) way, providing optimal provisioning of SIP-based multimedia services according to user preferences.

Keywords: context-aware service provisioning, IMS-enabled services, session mobility, multi-agent system.

1 Introduction

The advent of the Internet and the development of the Next-Generation Network (NGN) is enabling a lifestyle aspiring to digital humanism where people's daily activities are becoming more digitalized, convenient and intelligent [1]. With consumers typically having several multi-purpose end-user devices, the number and variety of personal, work, and home related services offered will also grow. In order to increase revenue, actors on telecom markets are pursuing innovations and

R. Kowalczyk et al. (Eds.): SOCASE 2009, LNCS 5907, pp. 71–82, 2009.

launching new value-added services (VAS) [2]. This new lifestyle and use of VAS created a whole new market demand and opened various business opportunities in a global multi-service and multi-provider market. This new market demand and technological development has led to the convergence of different domains (i.e., telecommunications, information technology (IT), the Internet, broadcasting and media), all involved in the telecom service provisioning process.

The ability to transfer information embodied in different media into digital form to be deployed across multiple technologies is considered to be the most fundamental enabler of convergence [3]. The realization of the full potential of convergence will make it necessary for operators to deploy a dynamic, cooperative and business-aware consistent knowledge layer in the network architecture in order to enable ubiquitous personalized services. The evolved network should aim at taking changing customer demands into account and creating spontaneous, adaptive services that can be delivered anytime, anywhere, to any device the user prefers. Providing such context-aware services transparently to the user is not only challenging from a network point of view, but also places severe requirements on service provisioning. This is particularly true when the service is to be accessed across several administrative domains (i.e., in a multi-provider environment).

The number of telecom value-added service consumers is rising continuously. Moreover, the competition among stakeholders in the telecom market, as well as the NGN concept which introduces a whole new spectrum of services, enables consumers to be very picky. Consequently, realization of the full potential of the NGN will make it necessary for service providers to offer dynamic, ubiquitous and context-aware personalized services. Moreover, a large part of NGN services will provide multimedia sessions which will be composed of different audio and/or video communications with a certain quality of service (QoS) [4]. Providing such services to consumers transparently is challenging from the technical, business and social points of view.

The rest of the paper is structured as follows. Section 2 addresses the position of our multi-agent system in the telecommunications environment it is placed in. Furthermore, it gives an overview of the main technological foundations used in our proof-of-concept prototype which enables advanced service provisioning options in the NGN. Section 3 describes the architecture and the main features of our multi-agent system, while Section 4 presents the agent-based personalized session mobility service. Section 5 concludes the paper and gives an outline for future work.

2   Technological Foundations

Our proof-of-concept prototype is placed in a telecommunications environment and it is based on the following technological foundations presented in this section: the IP Multimedia System (IMS), the Session Initiation Protocol (SIP), the Java Agent DEvelopment Framework (JADE), the Java Expert System Shell (JESS) and Asterisk. In the designed multi-agent system, intelligent software agents use different technologies (e.g., IMS, JADE, JESS, Asterisk) to enable personalized service provisioning, automated coordination of network operator's operations and automated interaction between all entities in an NGN environment.

2.1   Telecommunications Environment

Today we are witnessing the fusion of the Internet and mobile networks into a single, but extremely prominent and globally ubiquitous, technology: the Network [5]. The Network will enable the transformation of physical spaces into computationally active and intelligent environments [6], characterized by ambient intelligence where devices embedded in the environment provide seamless connectivity and services at all times. The vision of the Network is becoming a reality with the new generation of communication systems: the NGN [7].

One of the fundamental principals in the NGN is the separation of services from transport [8]. This separation represents a horizontal relationship in the NGN where the transport stratum and the service stratum can be distinguished. The transport stratum encompasses the technical processes that enable three types of connectivity: user-to-user, user-to-service platform and service platform-to-service platform connectivity. On the other hand, the service stratum is comprised of business processes that enable (advanced) telecom service provisioning assuming that the earlier stated types of connectivity already exist. Each stratum can have multiple layers, representing vertical relationships, where each layer can be distinguished into a data (or user) plane, a control plane and a management plane. In our model, we introduce intelligent agents into the control plane and the management plane of the service stratum. These agents are in charge of gathering context information that is required for service personalization (i.e., service management) and facilitating personalized application-level mobility (i.e., service control).

2.2   IP Multimedia System

The first step in seamless provisioning of personalized ubiquities services is enabled by deployment of the IP Multimedia System (IMS) with the aim of offering IP-based real-time multimedia services. IMS enables the convergence of mobile and wireline networks into a single unified infrastructure in all its forms by supporting services independent of access. It has a layered structure comprised of 1) a service layer, 2) a control layer, and 3) a connectivity layer. The service layer consists of application and content servers which execute value-added services for a user. The control layer is composed of network control servers for managing call or session set-up, modification and release. The most important is the Call Session Control Function (CSCF). This layer also contains a full suite of support functions, such as provisioning, charging and operations and maintenance (O&M). Interworking with other operators' networks and/or other types of networks is handled by border gateways. The connectivity layer is comprised of routers and switches, both for the backbone and for the access network. IMS takes the concept of layered architecture one step further by defining a horizontal architecture where service enablers and common functions can be reused for multiple applications.

With IMS, personalization of user services are achieved via a dynamically associated, user-centric, service independent and standardized access point, the CSCF. The service architecture is user-centric and is highly scalable. However, IMS does not have a common framework for context-awareness across all functions in the control layer in order to automatically adapt service availability and delivery to heterogeneous networks and dynamically changing environments. In order to enable seamless provisioning of personalized ubiquities services in a converged telecom network, we propose an agent-based common framework for context-awareness.

2.3   Session Initiation Protocol

By applying SIP for call/session control in the mobile Internet, personalized ubiquitous services can incorporate all types of media, from continuous media to application sharing. Besides support for multimedia information, SIP enables seamless integration of user devices such as 3G mobile phones and laptops. Moreover, SIP enables integration of devices with resources embedded in the users environment, such as video projectors, video cameras, and loudspeakers. Active multimedia sessions can be moved from one device to another or can be split across devices [9,10].

In addition to mobility support, SIP enables deployment of context-aware services by utilizing its event mechanism extension which scales to a large number of users spread across different administrative domains. SIP event mechanism can be used for exchange of context-related information such as a user's current location, device capabilities, network characteristics, a user's interests, presence, time of day, etc. It is important that the exchange of context information is as privacy-conscious as possible (i.e., users should control the which part of their context information is revealed to others). Furthermore, wherever possible, context-aware services should take advantage of user defined policies, rather than requiring direct user interaction. For example, policies can be triggered dynamically by presence and location information. By using agent technology and SIP, multimedia communication can be seamlessly provisioned, taking into account not only mobility issues and context-awareness, but also the semantics of exchanged events. This can be accomplished through the introduction of formally (i.e., ontology-based) described metadata representing SIP events which enables intelligent agents to understand context information and dynamically trigger provisioning operations. Therefore, we propose building intelligent agents (Personal Assistants) that will enable users to consume context-aware ubiquitous services in a seamless way, providing enhanced provisioning of SIP-based multimedia services according to user preferences.

2.4   Java Agent DEvelopment Framework

The proposed multi-agent system was implemented using the Java Agent DEvelopment Framework (JADE[1]). JADE is a software framework used for developing

---
[1] http://jade.tilab.com

agent-based applications in compliance with the Foundation for Intelligent Physical Agents (FIPA[2]) specifications. These specifications define standard agent interaction protocols and other key aspects of a multi-agent system allowing interaction between different agent platforms and software components.

JADE agents are identified by a unique name (Agent IDentifier - AID). Assuming that the agents know each others names, they can communicate transparently regardless of their actual location. Communication is performed using Agent Communication Language (ACL) messages which agents exchange between themselves. The architecture is based on two agents that are initiated each time the platform is started - the Agent Management Service (AMS) agent and the Directory Facilitator (DF) agent. The AMS agent exerts supervisory control over access and use of the platform. The DF agent provides a Yellow Pages service which enables agents to find other agents providing the services they requires in order to achieve their goals. The third component of the platform is the Agent Communication Channel (ACC). The ACC is a Message Transport System that controls the exchange of all messages within the platform, including messages to/from remote platforms.

2.5   Java Expert System Shell

Rule-based programming is appropriate for problems that are difficult to solve using traditional algorithmic methods [11]. Consequently, it has to be executed by a of run-time system that understands how to control its flow and how to use declarative information to solve problems. A rule-based program does not consist of one long sequence of instructions; instead, it is made up of discrete rules, each of which applies to a subset of the problem. A rule engine determines which rules apply at any given time and executes them accordingly.

Our service provisioning solution has adopted the Java Expert System Shell (JESS[3]) as both a rule engine and a scripting language for the specification of rules. JESS was developed at Sandia National Laboratories in Livermore, California in late 1990s [11]. Using Jess, Java software can be built with the capacity to "reason" using knowledge that the programmer supplies in the form of declarative rules. JESS rules are very similar to 'if-then' statements in traditional programming languages. The rule-based system uses rules to reach conclusions from a set of premises. A JESS rule-based system consists of a working memory, an inference engine and a rule base. The rule base contains all the rules that the system knows, while the working memory contains all the facts that the system works with. The inference engine has three components: a pattern matcher, an agenda and an execution engine. The pattern matcher decides which rules should be activated during the current cycle by comparing all of them with the facts currently present in the working memory. This list of rules is then stored in the agenda. The execution engine executes the first rule from the list, along with adding removing or modifying existing facts in the working memory if necessary. The entire process is then repeated.

---

[2] http://www.fipa.org

[3] http://herzberg.ca.sandia.gov/jess/

2.6 Asterisk

Asterisk[4] is an open source telephony engine and toolkit. It was originally created in 1999 and it represents an implementation of a telephone Private Branch eXchange (PBX[5]). Telephones connected to Asterisk can access the Public Switched Telephone Network (PSTN) and Voice over Internet Protocol (VoIP) services. It provides SIP and PSTN-based phones with call and session functionalities. Every SIP phone can register through it and can use it to call other SIP and PSTN phones. Asterisk also provides handlers for defining presence and session information since it can send event messages to components in the system when such events occur. In this paper, we consider the following two compliant Asterisk interfaces which we use in our prototype: The Asterisk Management Interface (AMI) which allows our multi-agent system to control and monitor the Asterisk system; and the Asterisk Gateway Interface (AGI) which allows our multi-agent system to control the Asterisk dial plan.

3 A-STORM Multi-agent System

The Agent-based Service and Telecom Operations Management (A-STORM) multi-agent system is part of the prototype that deals with agent-based service provisioning. The prototype has been developed in order to explore the possibilities of implementing ontology-based user profiling/clustering, context-aware service personalization and rule-based software deployment in the 3G mobile network. Figure 1 shows the implemented multi-agent system in the proof-of-concept prototype. The Business Manager Agent and the Provisioning Manager Agent belong to the group of business-driven provisioning agents whose task is to perform provisioning operations according to business strategies defined in a rule-based system. These strategies take into account business related information (e.g., user categories, service tariffs, season period, location of service execution, type of service content, business-to-business (B2B) relationships, etc.).

The Deployment Coordinator Agent and Remote Maintenance Shell agents can be categorized as service deployment agents that provide end-to-end solutions for efficient service deployment by enabling software deployment and maintenance at remote systems. Moreover, they provide so-called software component mobility (i.e., software components can seamlessly be relocated from one network node to another).

The Charging Manager Agent, Group Manager Agent, Session Manager Agent and Preference Manager Agent form a group of context management agents. They gather context information from network nodes and terminals (e.g., triggered events in SIP/PSTN call model, balance status, terminal location) and enable user personalization through the execution of context-dependent personal rules.

---

[4] http://www.asterisk.org
[5] http://en.wikipedia.org/wiki/Asterisk_(PBX)

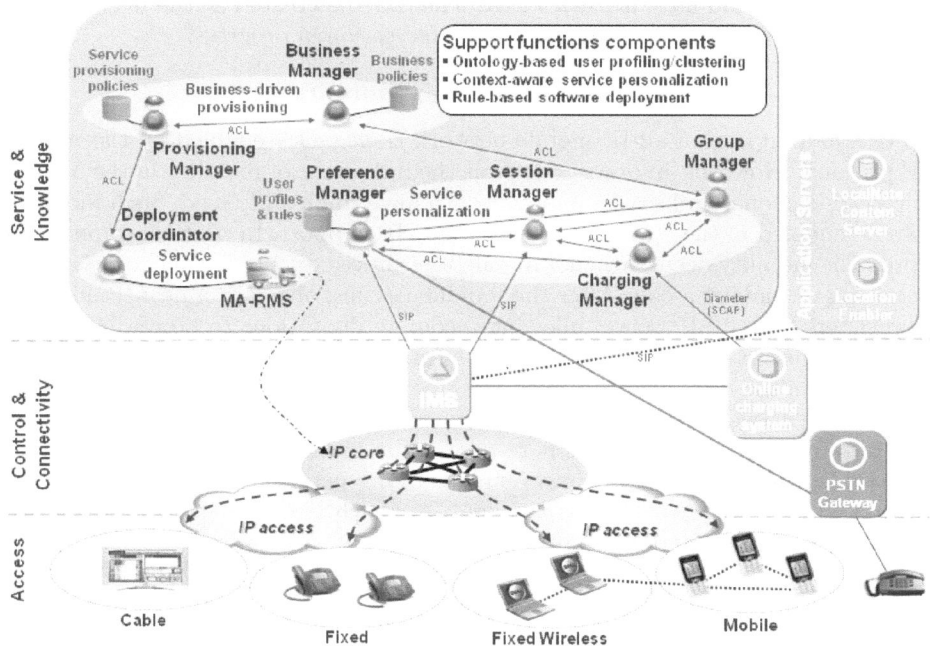

Fig. 1. A-STORM proof-of-concept prototype architecture

The Business Manager Agent, Charging Manager Agent, Provisioning Manager Agent, Deployment Coordinator Agent and Remote Management Shell Agents' tasks are beyond the scope of this paper. For a more detailed description of these agents and their functionalities can be found in [12,13,14,15,16,17]. Their features are prerequisites for context-aware provisioning of IMS-enabled ubiquitous services. Two agents (the Preference Manager Agent (PMA) and the Session Manager Agent (SMA)) discussed later in this paper are incorporated in the prototype in order to facilitate personalized application-level mobility features supported by SIP. Furthermore, they provide semantic interoperability of context information delivered by the SIP event mechanism.

The context information that is required for service personalization is managed by the PMA which follows personal rules specified by the user it represents. The personal rules depend on context information (e.g., user location, session/balance status, presence) gathered from network nodes that are, for example, part of an Online Charging System[6] or an IMS, enhanced with a location enabler and a media gateway towards traditional PSTN terminals. Moreover, the PMA handles the knowledge base which contains profiles of user preferences and terminal capabilities, enabling service personalization in different types of user devices. Each user

---

[6] http://www.ericsson.com/mobilityworld/sub/open/technologies/
charging_solutions/tools/diameter_charging_sdk

in the NGN should have its own PMA. The SMA is created at the beginning of the session and is in charge of monitoring the session in progress.

3.1   Agent-Based Support for Session Mobility

Device mobility in an all-IP mobile network is elegantly enabled by the mobile IP protocol. However, in order to exploit the full power of mobility in the Mobile Internet, personal and session mobility also have to be addressed. Such mobility can be enabled in the application level with SIP support. In particular, personal (pre-call) mobility occurs when a mobile user moves to another network prior to receiving or making a call. After the mobile user has obtained a new IP address, it registers with a SIP server allowing incoming invitations to be re-directed to the mobile user's current location. Session (mid-call) mobility, on the other hand, occurs when the user changes a terminal, moves to another network, or switches to another network interface during an ongoing session. After the mobile user has obtained a new IP address it re-invites the correspondent host in order to re-establish communication. However, addressing personal and session mobility in a dynamic heterogeneous environment, in which both resources and services vary in terms of availability and configuration, is a real challenge.

3.2   SIP-Based Knowledge Exchange of Service Context

Context-awareness refers to the ability to detect and incorporate information regarding the user, network conditions, terminal capabilities, etc. The SIP event mechanism represents an elegant and scalable solution for the delivery of all types of context information. This information can be spread over different devices, can involve several pieces of service logic, and may reside in several administrative domains. However, the SIP event mechanism does not provide any application-level semantic description of the delivered events between distributed parties. We propose a solution for semantic interoperability by building an agent-based knowledge exchange system that will enable the sharing and unambiguous understanding of context-related information through the use of ontologies.

The intelligent agents used in our prototype are equipped with reasoning capabilities enabling them to understand information delivered to them by the SIP event mechanism. Namely, they are able to understand ontology-based context information by utilizing a description logic (DL)-based reasoning engine. Consequently, there is a high degree of automation in knowledge exchange between software agents in the prototype. The agents (e.g., the GMA and the PMA) use OWLS-MX [18], a hybrid semantic matching tool which combines logic-based reasoning with approximate matching, based on syntactic IR similarity computations. The GMA uses this for semantic clustering of users, while the PMA uses it for the discovery of eligible services according to user preferences.

4   Case Study

Figure 2 shows entities included in the case study described in this section. All entities communicate with each other by exchanging ACL or SIP messages. The

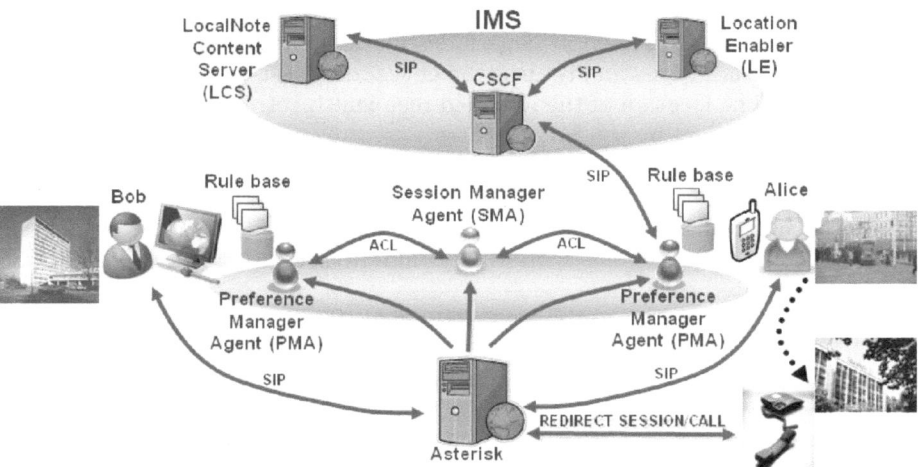

Fig. 2. Session mobility

LocalNote service [19] can be described as a location-triggered instant messaging service. It provides a mechanism for sending short text messages whereby the sender can specify the area in which the recipient must reside in order to receive the message. The LocalNote Content Server (LCS) is the core server of the LocalNote service and is used to trigger redirection of the established SIP session. It contacts the IMS enabler, known as the Location Enabler (LE), which is used to obtain information regarding recipients' positions. The PMA subscribes to a user's location, while the LE notifies the PMA when the user enters the area specified in his subscription.

4.1   SIP Entities

In this subsection, we describe the SIP elements that are used in our proof-of-concept prototype. Basic SIP elements present in a typical SIP network are: User Agents, SIP Proxy Servers, SIP Registrars and Redirect Servers.

User Agents are Internet end points that use SIP to find each other and negotiate session characteristics. They usually reside on users' computers in the form of applications, but can also reside in mobile phones, PSTN gateways, PDAs and so on. In our prototype, we consider two users: Alice and Bob. Bob's User Agent is on his office desktop PC, while Alice has two User Agents: one on her mobile phone and the other on her office PSTN phone.

SIP allows creation of an infrastructure of network hosts called SIP Proxy Servers. They are important entities in the SIP infrastructure because they perform routing of session invitations according to invitee's current location, authentication, accounting and many other important functions. The most important task of a Proxy Server is to route session invitations "closer" to the callee.

Redirect Servers respond to a SIP request with an address where the SIP message should be redirected. It maps a destination address (in the SIP message) to one or more addresses and returns the new address list to the originator of the SIP request. The location of the intended recipient is retrieved from the location database maintained by the SIP Registrar.

The Registrar is a special SIP entity that receives registrations from users, extracts information regarding their current location and stores this information in a location database. The location database is then used by the Proxy Servers. Due to their tight coupling with Proxy and Redirect Servers, Registrars are usually co-located with them.

Proxy Servers are referred to as CSCFs [20]. We distinguish between Proxy CSCFs (P-CSCF), Serving CSCFs (S-CSCF), and Interrogating CSCFs (I-CSCF). A P-CSCF represents the point of contact between the user terminal and the rest of the network. A S-CSCF provides services to the user while an I-CSCF's role is to find the proper S-CSCF for a particular user.

In our proof-of-concept prototype we used the Ericsson Service Development Studio[7] to simulate an IMS environment and run Application Servers (e.g., PCS, LE).

4.2   Session Mobility

In order to make or receive a SIP call, the user must register, after which Asterisk informs the user's PMA about the registration. The PMA updates the user's rule engine with information that the user is available.

User Bob initiates a SIP session with user Alice. Asterisk establishes the session and creates the SMA in charge of the created session. Bob is connected with his desktop PC, while Alice is in town talking to Bob from her mobile phone. The SMA informs Alice's and Bob's PMAs, which must update their databases. When Alice's PMA updates her database with the information that she is in an ongoing session on her mobile phone, a rule is triggered. This rule states that when Alice enters her office talking on her mobile phone, the session should be redirected to her office PSTN phone. Consequently, the PMA sends a message to the LE, requesting notification when Alice enters her office. The LCS subscribes to information regarding Alice's location with the LE.

During an ongoing session, Alice arrives to her office. The LE sends a notification message to the LCS, which in turn forwards this message to Alice's PMA. Her new location is written in her database, triggering another rule which ensures that the PMA informs the SMA that the session should be redirected. The SMA then sends a redirect request to Asterisk which initiates redirection of the ongoing session from Alice's mobile phone to her office PSTN phone. After the session has been redirected to the PSTN phone, the mobile phone is excluded from the rest of the session.

---

[7] http://www.ericsson.com/mobilityworld/sub/open/technologies/ims_poc/tools/sds_40

5   Conclusion and Future Work

Agent technology presents a promising solution for service provisioning problems associated with 3G mobile networks. The A-STORM multi-agent system, which consists of intelligent and mobile agents, provides possibilities for implementing proactive optimized service deployment, charging-aware service provisioning, and support for multi-provider environments in the 3G mobile network. We propose a multi-agent architecture which, in conjunction with SIP, can realize the mobility of SIP-based ubiquitous services. The functionalities of two agents (the Preference Manager Agent (PMA) and the Session Manager Agent (SMA)) are presented. The PMA follows personal rules, specified by its user, and manages the context information that is required for service personalization. After a session between two users has been established, the SMA is created and is in charge of monitoring and managing the ongoing session. In cooperation with SIP entities, these two types of agents enable personalized session mobility.

Further development of our multi-agent system is aimed at enabling context-aware provisioning of group-oriented services that will use context information, not only in IMS-based networks, but also in emerging sensor networks.

Acknowledgments. The work presented in this paper was carried out within research projects 036-0362027-1639 "Content Delivery and Mobility of Users and Services in New Generation Networks", supported by the Ministry of Science, Education and Sports of the Republic of Croatia, and "Agent-based Service & Telecom Operations Management", supported by Ericsson Nikola Tesla, Croatia.

References

1. Yoon, J.L.: Telco 2.0: a new role and business model. IEEE Communications Magazine 45(1), 10–12 (2007)
2. Damsgaard, J., Marchegiani, L.: Like Rome, a mobile operator's empire wasn't built in a day!: a journey through the rise and fall of mobile network operators. In: Janssen, M., Sol, H.G., Wagenaar, R.W. (eds.) ICEC. ACM International Conference Proceeding Series, vol. 60, pp. 639–648. ACM, New York (2004)
3. Hanrahan, H.: Network Convergence: Services, Applications, Transport, and Operations Support. John Wiley & Sons, Inc., New York (2007)
4. Ghys, F., Vaaraniemi, A.: Component-based charging in a next-generation multimedia network. IEEE Communications Magazine 41(1), 99–102 (2003)
5. Podobnik, V., Petric, A., Trzec, K., Jezic, G.: Software agents in new generation networks: Towards the automation of telecom processes. In: Jain, L.C., Nguyen, N.T. (eds.) Knowledge Processing and Decision Making in Agent-Based Systems, pp. 71–99. Springer, Heidelberg (2009)
6. Weiser, J.S.: The coming age of calm technology. In: Denning, R. (ed.) Beyond Calculation: The Next Fifty Years of Computing, Copernicus (1998)
7. Ljubi, I., Podobnik, V., Jezic, G.: Cooperative mobile agents for automation of service provisioning: A telecom innovation. In: ICDIM, pp. 817–822. IEEE, Los Alamitos (2007)

8. General principles and general reference model for next generation networks. Technical Report ITU-T Recommendation Y.2011, Telecommunication Standardization Sector (2004)

9. Schulzrinne, H., Wu, X., Sidiroglou, S., Berger, S.: Ubiquitous computing in home networks. IEEE Communications Magazine 41(11), 128–135 (2003)

10. Berger, S., Schulzrinne, H., Sidiroglou, S., Wu, X.: Ubiquitous computing using SIP. In: Papadopoulos, C., Almeroth, K.C. (eds.) NOSSDAV, pp. 82–89. ACM, New York (2003)

11. Hill, E.F.: Jess in Action: Java Rule-Based Systems. Manning Publications Co., Greenwich (2003)

12. Petric, A., Ljubi, I., Trzec, K., Jezic, G., Kusek, M., Podobnik, V., Jurasovic, K.: An agent based system for business-driven service provisioning. In: O'Sullivan, B., Orsvarn, K. (eds.) Proceedings of the AAAI 2007 Workshop on Configuration, pp. 25–30. AAAI Press, Menlo Park (2007)

13. Vrancic, A., Jurasovic, K., Kusek, M., Jezic, G., Trzec, K.: Service provisioning in telecommunication networks using software agents and rule-based approach. In: Proceedings of the 30th International Convention MIPRO 2007, pp. 159–164 (2007)

14. Dumic, G., Podobnik, V., Jezic, G., Trzec, K., Petric, A.: An agent-based optimization of service fulfillment in next-generation telecommunication systems. In: Car, Z., Kusek, M. (eds.) Proceedings of the 9th International Conference on Telecommunications ConTEL 2007, Zagreb, Croatia, pp. 57–63 (2007)

15. Grubisic, D., Kljaic, K., Ljubi, I., Petric, A., Jezic, G.: An agent-based user selection method for business-aware service provisioning. In: Proceedings of the 15th International Conference on Software, Telecommunications and Computer Networks, Split, Croatia, pp. 1–5 (2007)

16. Jezic, G., Kusek, M., Desic, S., Labor, O., Caric, A., Huljenic, D.: Multi-agent system for remote software operations. In: Palade, V., Howlett, R.J., Jain, L.C. (eds.) KES 2003. LNCS, vol. 2774, pp. 675–682. Springer, Heidelberg (2003)

17. Lovrek, I., Jezic, G., Kusek, M., Ljubi, I., Caric, A., Huljenic, D., Desic, S., Labor, O.: Improving software maintenance by using agent-based remote maintenance shell. In: ICSM, pp. 440–449. IEEE Computer Society, Los Alamitos (2003)

18. Klusch, M., Fries, B., Sycara, K.P.: Automated semantic web service discovery with OWLS-MX. In: Nakashima, H., Wellman, M.P., Weiss, G., Stone, P. (eds.) AAMAS, pp. 915–922. ACM, New York (2006)

19. Brajdic, A., Lapcevic, O., Matijasevic, M., Mosmondor, M.: Service composition in IMS: A location based service example. In: 3rd Int. Symposium on Wireless Pervasive Computing, ISWPC 2008, Santorini, Greese, pp. 208–212 (2008)

20. Sinnreich, H., Johnston, A.B.: Internet communications using SIP: delivering VoIP and multimedia services with Session Initiation Protocol. John Wiley & Sons, Inc., New York (2001)

Agent-Based Framework for Personalized Service Provisioning in Converged IP Networks

Vedran Podobnik[1], Maja Matijasevic[1], Ignac Lovrek[1], Lea Skorin-Kapov[2], and Sasa Desic[2]

[1] University of Zagreb, Faculty of Electrical Engineering and Computing, Croatia
[2] Ericsson Nikola Tesla, R&D Center, Croatia
{vedran.podobnik,maja.matijasevic,ignac.lovrek}@fer.hr,
{lea.skorin-kapov,sasa.desic}@ericsson.com

Abstract. In a global multi-service and multi-provider market, the Internet Service Providers will increasingly need to differentiate in the service quality they offer and base their operation on new, consumer-centric business models. In this paper, we propose an agent-based framework for the Business-to-Consumer (B2C) electronic market, comprising the Consumer Agents, Broker Agents and Content Agents, which enable Internet consumers to select a content provider in an automated manner. We also discuss how to dynamically allocate network resources to provide end-to-end Quality of Service (QoS) for a given consumer and content provider.

Keywords: agent-based B2C e-market; Quality of Service; Internet business environment; provider selection.

1 Introduction

Internet service providers (ISPs) and IP-based telecom network operators are turning towards new business opportunities in a global multi-service and multi-provider market. With consumers typically having several multi-purpose end-user devices, the number and variety of personal, work, and home related services offered will also grow. As "plain broadband" wired/wireless Internet access is likely to become a commodity in the next 10 years or so [1], the ISPs will have to differentiate in the service quality they offer, and base their operation on new, consumer-centric business models. Such models involve a number of actors involved in service delivery, from the user (consumer of the service) to the end service provider, where the selection of the service provider is a non-trivial issue, considering an electronic market (e-market) with a number of service providers offering the same or similar service. The challenge is twofold: first, how to select "the best" service provider, given the user preferences and semantic service descriptions; and second, once the selection is made, how to dynamically allocate network resources on an end-to-end basis. The main contribution of this paper focuses on the first issue by proposing a novel agent-based framework for service provider selection. This process is based on service discovery which considers not only the semantic matching, but also the price and reputation of the service provider, in which it differs from other approaches found in literature. Section 2 gives

R. Kowalczyk et al. (Eds.): SOCASE 2009, LNCS 5907, pp. 83–94, 2009.

the problem formulation, while Section 3 presents the model. We discuss the second issue, end-to-end Quality of Service (QoS) in Section 4. Section 5 concludes the paper.

2 Roles and Relationships in the Electronic Market

There are a number of actors present in the Internet business environment who need to establish relationships in order to provide consumers with converged services. An actor may take on a number of roles in a particular scenario, and furthermore a number of actors can play the same role. The key roles and relationships, shown in Fig. 1, include *Consumer*, *Access Line Provider*, *Primary Service Provider* (PSP), *Internet Service Provider* (ISP), *Content Provider* (CP), and *Transport Provider* (TP) [2].

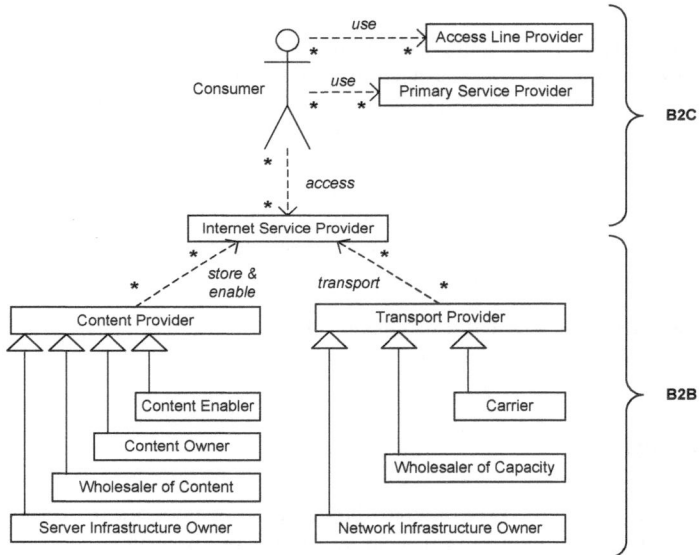

Fig. 1. Roles and relationships of actors in the Internet business environment

The *Consumer* is a role which typically represents the human user. The *Access Line Provider* is a role representing the owner of the access line. The ISP, in the most general sense, is a business entity providing a user with service(s). To differentiate between the responsibilities of an ISP which involve dealing with (e.g., multimedia) content, and those related to managing the transport of the content over the network infrastructure, we introduce the roles of CP and TP, respectively. The role of ISP as CP will be relevant for the first problem addressed in this paper – the selection of the service provider, while the role of TP will be relevant for the second issue – ensuring the end-to-end QoS. From now, we will refer to CP and TP, instead of just "ISP", to disambiguate roles. The PSP is an ISP which provides to a consumer the service of Internet access and consequently has a business relationship with that consumer. It may be noted that a particular consumer can have multiple PSPs, but only one PSP can be active at any one time. By adopting the "one-stop responsibility" concept [3], the PSP is also perceived as being responsible for coordinating the QoS negotiation and adaptation process, while further relying on the services of sub-providers in order to secure an end-to-end service and quality level to the consumer. Fig. 1 also shows the relationships between roles as Business-to-Customer (B2C) and Business-to-Business (B2B).

Our proposed agent based framework addresses the many-to-many relationship between Consumer and ISP in the B2C electronic market, as shown in Fig. 2. The roles are modeled by using an agent paradigm [4], as follows: 1) the *consumer representation* is represented by a Consumer Agent, 2) the *content provisioning* is represented by a Content Agent, and 3) the *brokering* between Consumer Agents and Content Agents is represented by a Broker Agent. The Broker Agent belonging to a certain ISP offers not only its own content (acting as a CP), but also the content offered (advertised) by other ISPs or CPs to which this ISP has established business relationships, as shown in Fig. 2.

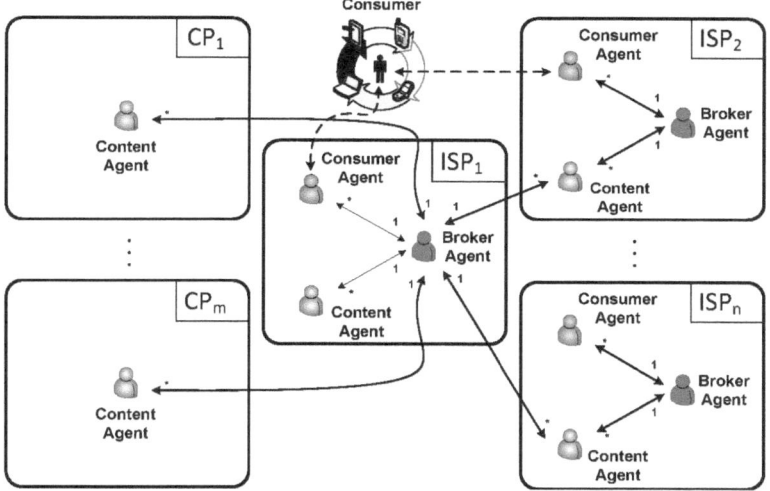

Fig. 2. The proposed agent framework for B2C electronic market

The issue of selecting a content provider has significant implications for all aspects of service provisioning. In the multi-provider network, a consumer will typically have a choice of a (possibly large) number of ISPs/CPs for a given content, as shown in Fig. 2. He or she may also have personal preferences regarding service options, device, and/or a particular wireline/wireless access network. We assume that the Consumer Agent containing these preferences is formed at the time of signing a contract with a PSP (e.g., fixed access via xDSL at home, and mobile access via HSPA in a 3G mobile network), and it resides within the PSP. Given all that, it is the task of the Broker Agent to discover the content and select the best match for the particular content request that corresponds to the given consumer preferences. It is also assumed that the scale of the problem is such that it cannot be solved by exhaustively querying all possible CPs.

As a running example we use the following problem, illustrated in Fig. 3: the user (Consumer) wants to view video-clips with Bayern goals from the latest Bundesliga round on her dual 3G/WLAN mobile phone. Her current (active) PSP is ISP_1. There are several CPs offering and advertising their service of providing *video clips of European football matches* (CP_i, ISP_j, ISP_k) to ISP_1. The ISP_1 has a business and technical relationship (SLA [3]) with ISP_j, and ISP_j has one with ISP_k. After receiving the Consumer request for content, the content from ISP_k is selected as the best match (the most eligible content). Then the QoS negotiation and adaptation takes place on end-to-end basis for a given consumer, service, and ISPs involved in service delivery, and having completed that, the service provisioning starts.

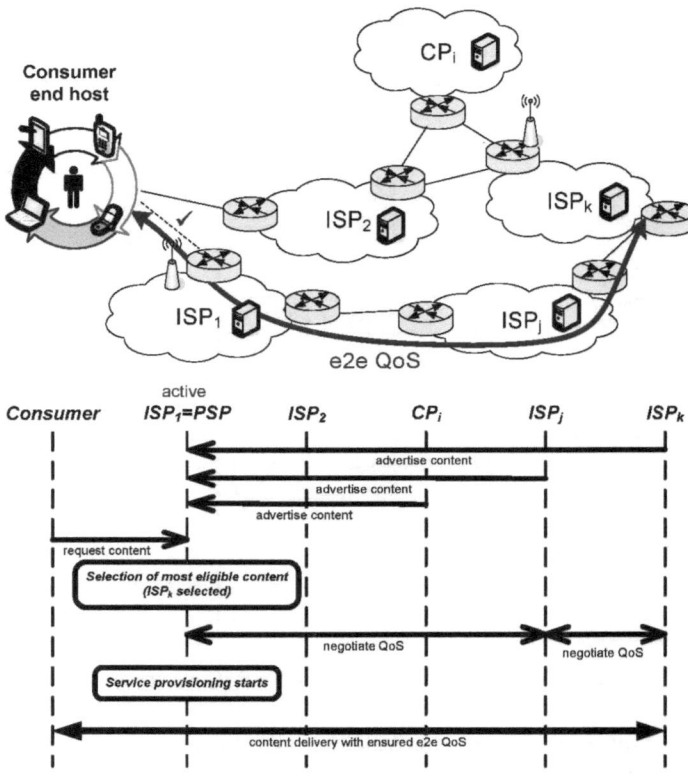

Fig. 3. Problem illustration by example

3 Selection of the Content Provider

The first step in the process of selecting the content provider is discovery. The state-of-the art discovery mechanisms are based on matching the semantics of resource (e.g., content) descriptions (i.e., semantic matchmaking) [5][6][7], rather than keyword matching [8]. The semantic dimension of resources such as multimedia content has been exploited in order to evaluate "interesting" inexact matches [9]. Most approaches suggested for semantic discovery use standard DL (*Description Logic*) reasoning to automatically determine whether one resource description matches the other. Our discovery mediator in the B2C e-market differs from previous approaches in that it considers the actual performance of businesses which act as ISPs/CPs (with respect to both price and reputation) in addition to semantic matchmaking. The mechanism on which the mediator is based is the Semantic Pay-Per-Click Agent (SPPCA) auction, a novel auction mechanism based on Pay-Per-Click (PPC) advertising auctions [8], but adapted for agent environment and enhanced with a semantic dimension [10].

3.1 The Architecture of Electronic Market for Content Trading

A description of the proposed agent-mediated B2C e-market architecture (Fig. 2) follows along with a demonstration of how it operates.

The Content Agent. In the proposed B2C e-market agents trade with various types of content (formally defined as a set $\mathcal{IC}$):

$$\mathcal{IC} = \{ic_1, ic_2, ..., ic_{|\mathcal{IC}|}\},$$

which is provided by different Content Providers (formally defined as a set $\mathcal{CP}$):

$$\mathcal{CP} = \{cp_1, cp_2, ..., cp_{|\mathcal{CP}|}\}.$$

Content Providers are represented in the e-market by Content Agents (formally defined as a set $\mathcal{A}_{\mathcal{CP}}$):

$$\mathcal{A}_{\mathcal{CP}} = \{a_{cp_1}, a_{cp_2}, ..., a_{cp_{|\mathcal{CP}|}}\}.$$

An a_{cp_i} represents a cp_i which offers a certain content ic_i that is described by content ontology, whose fragment (describing *video clips of European football matches*) is presented later in this work. Initially, a_{cp_i} wishes to advertise its content (advertised ic_i is denoted as ic_{adv}) at discovery mediator (i.e., the Broker Agent). An a_{cp_i} accomplishes that by participation in the SPPCA.

The Consumer Agent. Consumers of $\mathcal{IC}$ (formally defined as a set $\mathcal{C}$):

$$\mathcal{C} = \{c_1, c_2, ..., c_{|\mathcal{C}|}\},$$

are represented on the e-market by Consumer Agents (formally defined as a set $\mathcal{A}_{\mathcal{C}}$):

$$\mathcal{A}_{\mathcal{C}} = \{a_{c_1}, a_{c_2}, ..., a_{c_{|\mathcal{C}|}}\}.$$

An a_{c_i} acts on behalf of its human owner (i.e., consumer) in the discovery process of suitable ic_{adv} and subsequently negotiates the utilization of that content. An a_{c_i} wishes to get a best-ranked advertised content which is appropriate with respect to its needs (requested ic_i is denoted as ic_{req}).

The Broker Agent. Mediation between content requesters and content providers is performed by Broker Agents (formally defined as a set $\mathcal{A}_{\mathcal{B}}$):

$$\mathcal{A}_{\mathcal{B}} = \{a_{b_1}, a_{b_2}, ..., a_{b_{|\mathcal{B}|}}\}.$$

There is one a_{b_i} located at every ISP and it mediates between c (i.e., a_c) to whom this ISP is PSP and all cp (i.e., a_{cp}) which advertised its content at this a_{b_i}. An a_{b_i}

enables $\mathcal{A}_{C\mathcal{P}}$ to advertise their content descriptions and recommends the most eligible content to a_C in response to their requests. It is assumed that $a_{\mathcal{b}_i}$ is a trusted party which fairly mediates between content requesters and content providers.

3.2 The Content Discovery Model

Fig. 4 presents interactions between a_{c_i} and $a_{\mathcal{b}_i}$ which enable content discovery in the proposed B2C e-market. The a_{c_i}, by sending CFP (Call for Proposal) to $a_{\mathcal{b}_i}$, requests two-level filtering of advertised content descriptions to discover which is the most adequate for its needs. Along with the description of requested content ic_{req}, the CFP includes the set of matching parameters (to be explained later) that personalize the discovery process according to the consumer preferences. First-level filtering $(\mathcal{f}_1 : \mathcal{IC} \to \mathcal{IC})$ is based on semantic matchmaking between descriptions of content requested by c_i (i.e., a_{c_i}) and those advertised by cp (i.e., $a_{C\mathcal{P}}$). Content which passes the first level of filtering $(ic_{\mathcal{f}_1} \subset \mathcal{IC})$ is then considered in the second filtering step. Second-level filtering $(\mathcal{f}_2 : \mathcal{IC} \to \mathcal{IC})$ combines information regarding the actual performance of $cp_{\mathcal{f}_1}$ (cp which offer $ic_{\mathcal{f}_1}$) and prices bid in SPPCA by corresponding $a_{C\mathcal{P}_{\mathcal{f}_1}}$ ($a_{C\mathcal{P}}$ that represent cp which offer $ic_{\mathcal{f}_1}$). The performance of $cp_{\mathcal{f}_1}$ (with respect to both price and reputation) is calculated from the previous $\mathcal{A}_C$ feedback ratings. Following filtering, the most eligible content $(ic_{\mathcal{f}_2} \subset ic_{\mathcal{f}_1} : |ic_{\mathcal{f}_2}|=1)$ is chosen and recommended to the a_{c_i} in response to its request.

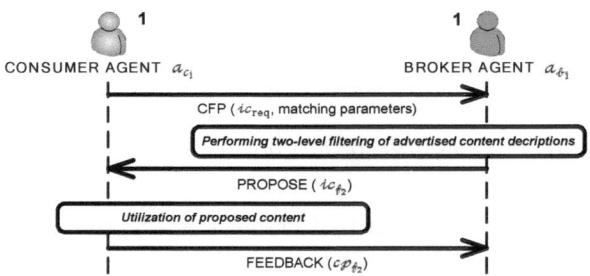

Fig. 4. The a_{c_i} discovers the most eligible content advertised at $a_{\mathcal{b}_i}$

Fig. 5 explains how the SPPCA auction, which is part of the discovery process, operates. The SPPCA auction is divided into rounds of fixed time duration. To announce the beginning of a new auction round, the $a_{\mathcal{b}_i}$ broadcasts a CFB (Call for Bid) message to all the $a_{C\mathcal{P}}$ which have registered their ic_{adv} for participation in the SPPCA auction. Every CFB message contains a status report. In such a report, the $a_{\mathcal{b}_i}$ sends to the a_{cp_i} information regarding events related to its advertisement which occurred during the previous auction round. The most important information is that regarding how much of the a_{cp_i} budget was spent (i.e., the advertisement bid price $\text{bid}_{ic_{\text{adv}}}$ multiplied by the number of recommendations of its ic_{adv} to various a_C). In response to a CFB message, an a_{cp_i} sends a BID message.

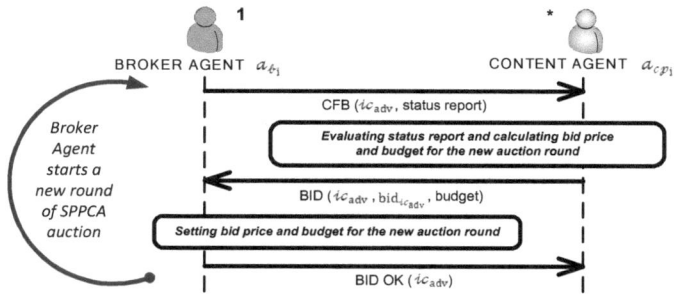

Fig. 5. The SPPCA auction

Semantic matchmaking of content descriptions. In the multi-agent system implementing the proposed B2C e-market model, the Semantic Web technology [11] is used to describe content. Namely, for describing content we use W3C's OWL-S (*Web Ontology Language for Services*), which is an OWL-based (*Web Ontology Language*) (Fig. 6) technology for describing the properties and capabilities of Web Services in an unambiguous, computer interpretable mark-up language.

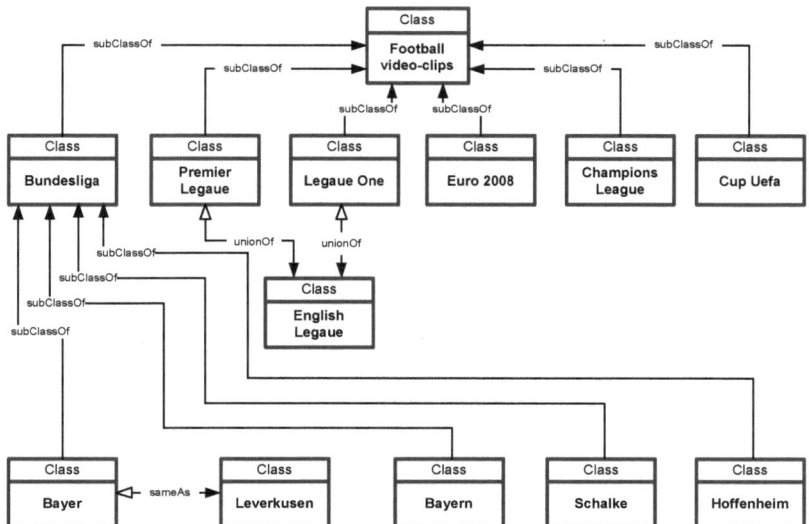

Fig. 6. The OWL ontology fragment describing *video clips of European football matches*

Our a_{b_i} uses OWLS-MX [12], a hybrid semantic matching tool which combines logic-based reasoning with approximate matching based on syntactic IR similarity computations. As the notion of match rankings is very important, OWLS-MX enables computation of the degree of similarity between compared service descriptions, i.e., the comparison is assigned a *content correspondence factor* (M). Namely, the OWLS-MX matchmaker takes as input the OWL-S description of a_{c_i} desired content ic_{req}, and returns a set of relevant content which match the query $ic_{\#_1}$. Relevant content is annotated with its individual degree of matching similarity value (i.e., $M_{ic_{req},ic_{adv}}$). There are six possible levels of matching [12]. The first level is a perfect match (also called an EXACT match) which is assigned a factor M = 5. Furthermore, we have four possible inexact match levels which are as follows: a PLUG-IN match (M = 4), a SUBSUMES match (M = 3), a SUBSUMES-BY match (M = 2) and a NEAREST-NEIGHBOUR match (M = 1). If two content descriptions do not match according to any of the above mentioned criteria, they are assigned a matching level of FAIL (M = 0). An a_{c_i} specifies its desired matching degree threshold (i.e., the M_{min}), defining how relaxed the semantic matching is.

The performance model of content providers. A performance model tracks the past performance of CP in the B2C e-market. Our model monitors two aspects of a cp_i performance – the reputation of the cp_i and the cost of utilizing the ic that cp_i is offering.

After utilizing the recommended content, an a_{c_i} gives an a_{θ_i} feedback regarding $cp_{\#_2}$, both from the reputation viewpoint (called the *reputation rating* (Q ∈ [0.0, 1.0])) and the cost viewpoint (called the *price rating* (P ∈ [0.0, 1.0])). A rating of 0.0 is the worst (i.e., the $cp_{\#_2}$ could not provide the content at all and/or utilizing the content is very expensive) while a rating of 1.0 is the best (i.e., the $cp_{\#_2}$ provides a content that perfectly corresponds to the c_i needs and/or utilizing the content is cost-efficient). The overall ratings of cp_i can be calculated in a number of ways. In our approach, we use the EWMA-based (*Exponentially Weighted Moving Average*) learning [13].

Calculating a recommended ranked set of eligible services. After an a_{θ_i} receives a CFP message from an a_{c_i} (Fig. 4), the discovery mediator finds the best-suitable content $ic_{\#_2}$ and recommends it to the a_{c_i} in response to its request. The final rating $R_{ic_{adv}}$ of a specific ic_{adv} at the end of discovery process is given by:

$$R_{ic_{adv}} = \frac{\alpha \times \dfrac{M_{ic_{req},ic_{adv}}}{5} + \beta \times Q_{cp_{adv}} + \gamma \times P_{cp_{adv}}}{\alpha + \beta + \gamma} \times bid_{ic_{adv}} \qquad (1)$$

A higher rating means that this particular ic_{adv} is more eligible for the consumer's needs (i.e., ic_{req}); α, β and γ are weight factors (i.e., matching parameters from CFP message in Fig. 4) which enable the a_{c_i} to personalize its request according to its owner's (i.e., $c_i's$) needs regarding the semantic similarity, reputation and price of a ic_{adv}, respectively; $M_{ic_{req},ic_{adv}}$ represents the *content correspondence factor* (M), but only ic_{adv} with M higher than threshold M_{min} are considered; $Q_{cp_{adv}}$ and $P_{cp_{adv}}$ represent the quality and price ratings of a particular cp_{adv}, respectively; $bid_{ic_{adv}}$ is the bid value for advertising an ic_{adv} in the SPPCA auction.

Since our performance model monitors two aspects of the cp_{adv} performance (i.e., its reputation and price), the $a_{c_{req}}$ defines two weight factors which determine the significance of each of the two aspects in the process of calculating the final proposal

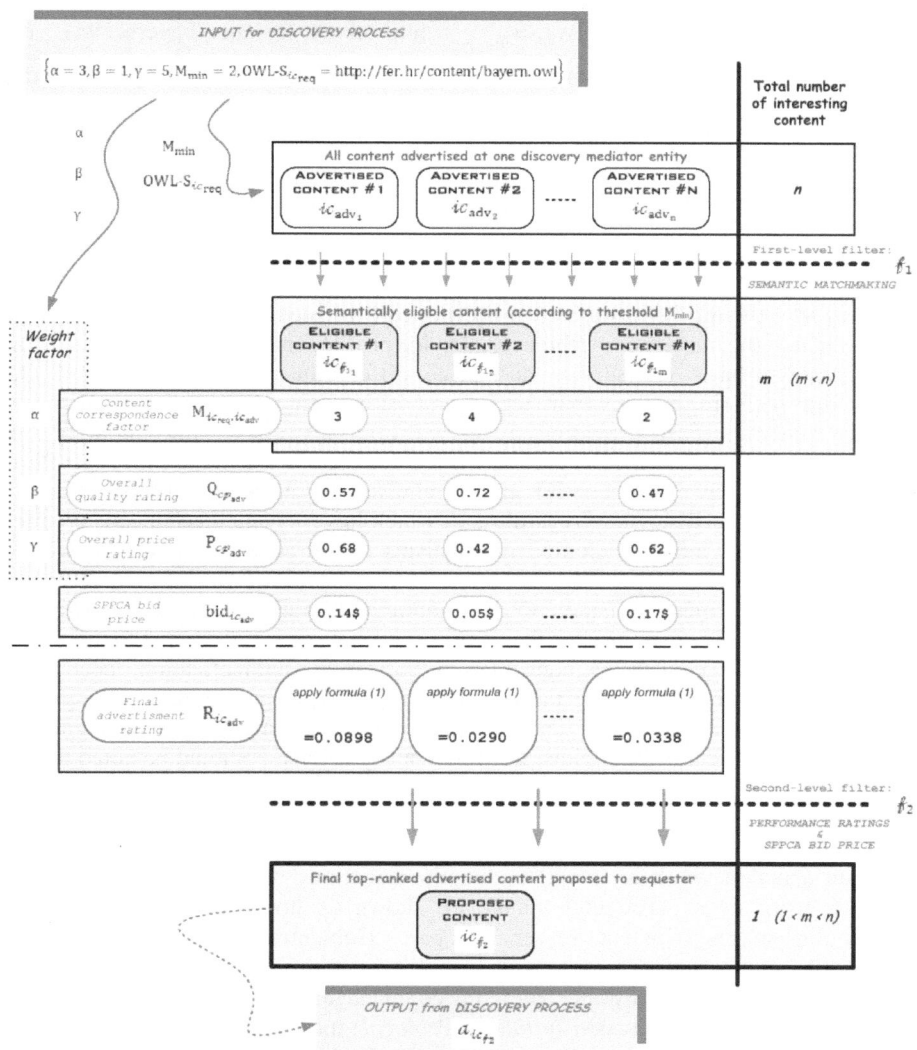

Fig. 7. An example of the discovery process

(β represents a weight factor describing the importance of cp_{adv} reputation while γ represents a weight factor describing the importance of content prices at cp_{adv}). Furthermore, an $a_{c_{req}}$ can specify whether information regarding the semantic similarity of ic_{req} and ic_{adv} is more important to it or information regarding an cp_{adv} performance. Thus, the $a_{c_{req}}$ also defines parameter α which is a weight factor representing the importance of the semantic similarity between ic_{ic} and ic_{adv}. In our example (Fig. 7) where requested content are video-clips with Bayern goals from the latest Bundesliga round $(\text{OWL-S}_{ic_{req}} = \text{http://fer.hr/content/bayern.owl})$, $\alpha = 3$, $\beta = 1$ and $\gamma = 5$. This means that the $a_{c_{req}}$ is looking for an inexpensive ic_{adv} and it is not very concerned with the cp_{adv} reputation.

4 Ensuring End-to-End QoS

Once the consumer, by using the mechanism described in the previous section, selected the CP, the end-to-end QoS needs to be negotiated with her PSP. For ensuring end-to-end QoS, support in the network is needed to negotiate and adapt QoS to match the consumer preferences, service profile, and network capabilities; and thus create a basis for service and price differentiation [14]. A general QoS negotiation scenario involves four steps: 1) a host initiating a service on another host; 2) the addressed host providing a service offer/answer; 3) the initiating host responding to the offer/answer; and 4) service delivery. We assume that all ISPs are QoS-aware, i.e., that they control and administer the necessary infrastructure for providing QoS-based services, regardless of which lower layers mechanisms are used. The selection of QoS provisioning mechanisms in the access and core network is performed in the TPs. Depending on the type of the network, this may involve various service control entities that handle QoS signaling, QoS policy control, and interaction with underlying network QoS mechanisms, as well as typical "support functions" (if and when needed), such as consumer authorization, authentication, accounting, auditing, and charging. During the process of QoS (re)negotiation, signaling flows typically traverse a number of functional network entities along the end-to-end path between communication endpoints, as shown in Fig. 8 [15]. It should be noted that the signaling (control) and data flows are separated, and that the resource managers in the data plane are "vertically" controlled by session control functions, which interface with the end-points and the internal databases related to consumers and services.

In an actual network architecture, functional entities may be mapped to one or more network nodes. The end points are shown as hosts (Host A, Host B) or application servers (Content Server, 3rd party Content/Application Servers). The additional functionality which must be implemented in the host may include, for example: a GUI for consumer preferences management, capability to negotiate session QoS (e.g., SIP (*Session Initiation Protocol*) interface), resource management capability (e.g., DiffServ), and mobility (e.g., Mobile IP's Mobile Host entity). Having in mind the CP selection procedure described in the previous section, we assume that a Consumer A, attached to the NGN by using Host A, has selected a PSP here shown as PSP Domain (Consumer A), to perform service control functions and offer access to 3rd party applications and services.

This service provider is then responsible for AAA functions (consumer authentication, authorization, and accounting), service provisioning, and maintaining a database for storing consumer-related data. It further interacts with an underlying network provider, for example, a 3G mobile network provider, which provides media connectivity functions. In a real life scenario, a single operator may take on multiple roles, including that of both a service and a network provider.

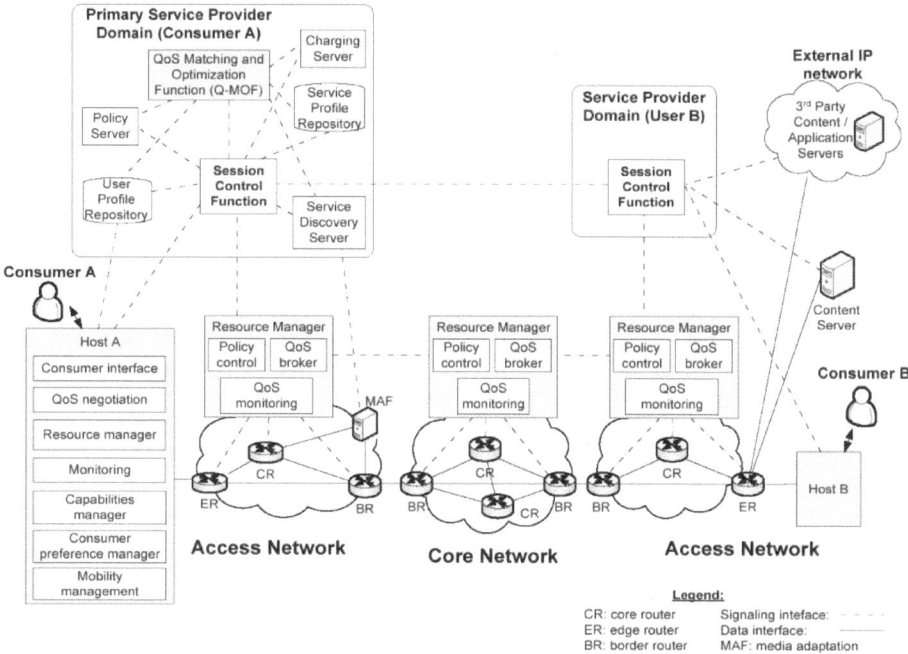

Fig. 8. End-to-End QoS provisioning

While the initial service matching and resource mapping may be based on QoS classes and SLA, more advanced mechanisms are needed to take into account dynamic changes in service profile (e.g., consumer willingness to pay for the service), network capabilities (e.g., due to handover), and service parameters (e.g., types of media streams comprising the service). Although the architecture proposed here is generic, in our previous work we considered a converged IP-based network based on 3GPP IP Multimedia Subsystem (IMS) [16]. In a pure-IP approach, SIP can be applied for session signaling, Diameter for policy control and AAA, and any QoS enabling mechanism may be applied at the network layer, including those for IP QoS interconnection [14][17][18].

5   Conclusion

In this paper, we proposed an agent-based framework for the B2C e-market where interactions between Consumer Agents, Broker Agents and Content Agents enable Internet consumers to select the most eligible content provider in an automated manner. The main benefit of the proposed approach is that in a situation with many ISPs/CPs offering the same or similar content, the user could not search for the content manually, nor exhaustively, nor could the best match be found based solely on semantic descriptions. Finally, we have discussed how end-to-end QoS could be negotiated once the content provider is selected.

Acknowledgements. The authors acknowledge the support of research project "Content Delivery and Mobility of Users and Services in New Generation Networks" (036-0362027-1639), funded by the Ministry of Science, Education and Sports of the Republic of Croatia, and projects "Agent-based Service & Telecom Operations Management" and "Future Advanced Multimedia Service Enablers" of Ericsson Nikola Tesla, Croatia.

References

1. Anderson, J.Q., Rainie, L.: The future of the Internet II. Technical report, Pew Internet and American Life Project (2006), http://www.pewinternet.org
2. Podobnik, V., Lovrek, I.: Multi-Agent System for Automation of B2C Processes in the Future Internet. In: 27th IEEE Conference on Computer Communications (INFOCOM) Workshops, pp. 1–4. IEEE Press, Phoenix (2008)
3. ITU-T Recommendation E.860: Framework of a service level agreement (2002)
4. Podobnik, V., Jezic, G., Trzec, K.: Towards New Generation of Mobile Communications: Discovery of Ubiquitous Resources. Electrotechnical Review 75(1-2), 31–36 (2008)
5. Sycara, K., Paolucci, M., Anolekar, A., Srinivasan, N.: Automated Discovery, Interaction and Composition of Semantic Web Services. Journal of Web Semantics 1(1) (2004)
6. Li, L., Horrock, I.: A Software Framework for Matchmaking Based on Semantic Web Technology. In: 12th International World Wide Web Conference (WWW), pp. 331–339. ACM, Budapest (2003)
7. Keller, U., Lara, R., Lausen, H., Polleres, A., Fensel, D.: Automatic location of services. In: Gómez-Pérez, A., Euzenat, J. (eds.) ESWC 2005. LNCS, vol. 3532, pp. 1–16. Springer, Heidelberg (2005)
8. Jansen, B.J.: Paid Search. Computer 39(7), 88–90 (2006)
9. Di Noia, T., Di Sciascio, E., Donini, F.M., Mong, M.: A System for Principled Matchmaking in an Electronic Marketplace. International Journal of Electronic Commerce 8(4), 9–37 (2004)
10. Podobnik, V., Trzec, K., Jezic, G.: Auction-Based Semantic Service Discovery Model for E-Commerce Applications. In: Meersman, R., Tari, Z., Herrero, P., et al. (eds.) OTM 2006 Workshops. LNCS, vol. 4277, pp. 97–106. Springer, Heidelberg (2006)
11. Leuf, B.: The Semantic Web: Crafting Infrastructure for Agency. John Wiley & Sons, New York (2006)
12. Klusch, M., Fries, B., Sycara, K.: Automated Semantic Web Service Discovery with OWLS-MX. In: 5th International Joint Conference on Autonomous Agents and Multiagent Systems (AAMAS), pp. 915–922. ACM, Hakodate (2006)
13. Luan, X.: Adaptive Middle Agent for Service Matching in the Semantic Web: A Quantitive Approach. PhD Thesis, University of Maryland (2004)
14. Briscoe, B., Rudkin, S.: Commercial models for IP quality of service interconnect. BT Technology Journal 23(2), 171–195 (2005)
15. Skorin-Kapov, L.: A framework for service-level end-to-end quality of service negotiation and adaptation. PhD Thesis, University of Zagreb (2007)
16. Skorin-Kapov, L., Mosmondor, M., Dobrijevic, O., Matijasevic, M.: Application-level QoS negotiation and signaling for advanced multimedia services in the IMS. IEEE Communications Magazine 45(7), 108–116 (2007)
17. Howarth, M.P., et al.: Provisioning for interdomain quality of service: the MESCAL approach. IEEE Communications Magazine 43(6), 129–137 (2005)
18. Masip-Bruin, X., et al.: The EuQoS System: A Solution for QoS Routing in Heterogeneous Networks. IEEE Communications Magazine 45(2), 9–103 (2007)

Management Intelligence in Service-Level Reconfiguration of Distributed Network Applications

K. Ravindran

City College of CUNY and Graduate Center,
Department of Computer Science,
160 Convent Avenue,
New York, NY 10031, USA
ravi@cs.ccny.cuny.edu

Abstract. The paper is on generic service-level management tools that enable the reconfiguration of a distributed network application whenever there are resource-level changes or failures in the underlying network sub-systems. A network service is provided to applications through a protocol module, with the latter exercising network infrastructure resources in a manner to meet the application-level QoS specs. Application requests for a network service instantiate the protocol module with parameters specified at the service interface level, along with a prescription of critical properties to be enforced. At run-time, a management module monitors the service compliance to application-prescribed requirements, and notifies the application if a QoS degradation is detected. In our model, the management entity uses policy scripts and rules to coordinate the application-level reconfigurations and adaptations in response to the changes/outages in underlying network services. Our management model is independent of the specifics of problem-domain, which lowers the software development costs of distributed applications through 'reuse' of the management module. The paper presents a case study of rate-adaptive video multicast over IP networks, to demonstrate the usefulness of our model.

1 Introduction

The management standards TINA and DCOM advocate the partitioning of a distributed network application into two types of entities: *service provider* (NSP) and *service user* (or client). The NSP maintains a repertoire of protocol mechanisms through which network services are offered to clients. For service delivery, a protocol mechanism exercises the infrastructure resource components placed at different sites of the network. A client application controls the extent to which network infrastructure resources are exercised, based on its QoS requirements. Our work on network service models purports to 'liaison' these two aspects of network application development.

R. Kowalczyk et al. (Eds.): SOCASE 2009, LNCS 5907, pp. 95–110, 2009.
© Springer-Verlag Berlin Heidelberg 2009

A service S may project the infrastructure resource availability onto a meta-state maintained by the underlying protocol $P(S)$ that exercises the infrastructure. The protocol state may in turn be mapped onto the service-level parameters visible to the clients of S. The service behavior determined therefrom allows clients to check on the compliance to service obligations required of S. See Figure 1. Any deviation from the expected critical requirements is symptomatic of service-level failures (e.g., frame miss rate seen by a video transfer application indicating the level of bandwidth congestion in the underlying network path). Thus, *behavioral monitoring* of network services is a necessary part of client-level reconfiguration mechanisms.

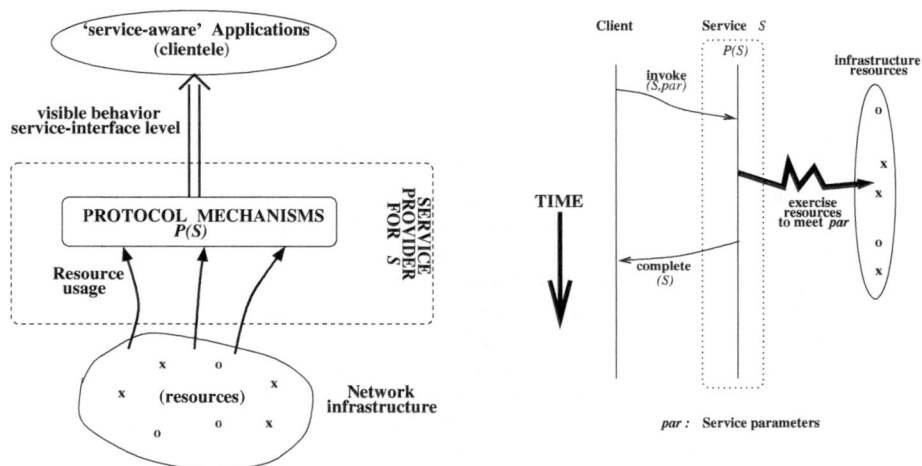

Fig. 1. Layers of functions in a network service

From a service-level programming standpoint, the monitoring and reconfiguration activities may be structured as management functions that can be instantiated with problem-specific parameters. For instance, how often the state variables characterizing a service behavior are sampled may be prescribed by the client agent, while the distributed mechanism to sample the state at different network nodes can itself be generic. Thus, a service-oriented management model can be employed across a variety of network applications to enable autonomic reconfigurations.

Our model of network service management, described in the paper, offers a generic template for developing a wide range of applications with reconfiguration capabilities. The associated signaling structures allow hooks into application-specific functions that carry out the client-level adaptations to service changes. The service-neutrality of our model arises from its 'functional' orientation, whereby a service prescription conforms to a generic template based on enumeration of service attributes and how they logically relate to one another in composing a desired service behavior. A management module inter-works with the service and client modules through generic interfaces, for the purpose of monitoring service compliance to client-prescribed critical properties. The monitor is realized by management-oriented software agents that can be implanted in the target client and service modules, and be projected onto the problem-specific parameter space to enable their inter-working with the target modules. A re-use of the agents across different applications can reduce the costs of developing distributed networking software. The paper presents a case study of network application: rate-controlled video multicast over IP networks, to demonstrate the usefulness our management model.

The paper is organized as follows. Section 2 provides a management view of adaptive network services. Section 3 presents the case study of a network application. Section 4 discusses related works. Section 5 concludes the paper.

2 Management View of Network Service Offering

The network service provider (NSP) provides a service S to a variety of distributed applications through an abstract *service interface* that prescribes a set of well-defined capabilities (or service features). An application may exercise one or more of these features to obtain a certain quality of service (QoS) through a protocol controlling the network infrastructure. For example, bounding the page access latency in a 'content distribution network (CDN) is a feature of 'content access' service that can be invoked by a web client by specifying an acceptable latency Δ as parameter. The CDN interface maps the Δ parameter onto a topological placement of proxy nodes for the master content server in the underlying content access protocol [1]. The latter then allows the client to reach a proxy node over a path with suitable latency characteristics. The choice of actual protocol mechanisms to provide a given level of service is itself hidden from the application.

2.1 Service Behaviors Visible to Applications

How the service 'quality' as seen by a client is affected by the changes in infrastructure resources constitutes a service behavior. Given a client-prescribed quality expected of a service S, a behavior captures — macroscopically — the exercising of infrastructure resources r by a protocol module $P(S)$ to deliver the required service. Protocol modules $P(S), P'(S), \cdots$ capable of providing S may exhibit distinct behavioral profiles, i.e., different levels of offering of S for a given amount of infrastructure resources r.

From a client's perspective, two services are identical if their externally visible behaviors for a given a set of parameters are the same. In the example of CDN, consider a dedicated content server for each client versus a shared content server across multiple clients. From a management standpoint, access latency with a low mean and variance is the performance benchmark. This view, while subsuming dedicated server based implementation of CDN (with implicit guarantees on latency), also encompasses shared server based implementations that exhibit a similar latency behavior — at least over an operating range of interest [1]. In general, service differentiation may be in terms of measurable parameters of externally visible behaviors.

Changes in a service behavior can be monitored in terms of measurable parameters of the service offering. Parameter changes depict macroscopic behavior indices (or symptoms) at the service interface, reflected onto from infrastructure changes. For example, a symptom of network congestion is the end-to-end data transfer delay over the network exceeding a limit (as in QoS-oriented network settings) [2,3]. In general, a service behavior can be dynamic in nature, i.e., service features can change at run-time due to changes in the underlying infrastructure elements.

2.2 Programmability of Service Offerings

With QoS-aware clients, a service degradation should be within prescribed tolerance limits. This richer notion of 'transparency' is captured by our network service model.

The client invokes a network service S by prescribing a set of parameters. The NSP instantiates an underlying network protocol $P(S)$ by mapping client prescriptions to a set of protocol-level parameters. The latter allow exercising the infrastructure resources by the functions that compose of $P(S)$. In general, a higher level of service obligation to clients requires an increased deployment of resources at the infrastructure level; likewise, a reduced resource availability lowers the QoS. In the example of CDN, the access latency and content accuracy client-specifiable parameters. They influence the placement of proxy nodes in the distribution topology and how the content in these nodes is dynamically updated: a lower latency requiring more proxy nodes and a higher accuracy requiring frequent updates of the proxy contents. The parameterizable execution of $P(S)$ allows a macro-level control of the infrastructure resource usage to meet the client-prescribed service needs. The relationship between resource usage and service quality is embodied into the functions that compose of $P(S)$.

Dynamically occurring infrastructure resource outages and/or failures (such as component upgrades or removals) may manifest as observable behavioral changes in a service offering. Based on a quantitative assessment of these changes, an *adaptive* client application may adjust its service parameters to match the availability of network infrastructure resources, at varying degrees. Even with sufficient infrastructure resources, a client may wish to relax its service expectations for many reasons: such as usage-sensitive service tariffs and time-varying profiles of service usage (e.g., longer access delay on content pages in a CDN may be acceptable at nights). In some other cases, an application may wish to increase its service expectations mid-point through a service provisioning (e.g., surplus bandwidth on a data path causing the sources to increase their send rates).

The afore-described service-level management can be seen through an ontological view of service offerings: DETECT ⟶ REACT ⟶ REPAIR cycle [4], executed at machine-level time-scales that are much faster than the application-perceivable time-scales. In a way, our model consciously allows a limited (and controlled) amount of damage but orchestrates corrections before these damages escalate into seriously higher level proportions. This is unlike the PREVENT paradigm which underlies the proactive methods of service protection in some existing approaches, say, by over-provisioning of resources.

2.3 Reactive Control of Network Applications

As systems become more and more complex (due to size, scale, heterogeneity, and diversity), a prevalent thinking among system designers is that it may be quite difficult to provide a 100% of QoS assurance, and it is rather prudent to learn to live with some limited damages to the system operations. This may require installing 'watch dogs' (i.e., service-level monitors) and recovery mechanisms at appropriate points in the control plane of the service architecture.

In the example of CDNs, the access latency on a content page is measured by noting the time elapsed from a transaction request to its completion (say, by time-stamping the client transactions). A large increase in latency indicates an excess queuing at the system resources (such as server disks and CPUs). This may in turn cause the management station to reduce the rate of transaction flows incident on the CDN service until the latency falls below the acceptable limit Δ.

Figure 2 shows our experimental studies on the time-scales of congestion detection at various layers of a video distribution network. In these studies, the human-perceived Qos degradation is on a slow time-scale, when compared to the agent-level detection based on the fluctuations in observed frame loss rate. Accordingly, a recovery triggered by user-level notifications (say, by a GUI 'callback' on sources) incurs higher latency, and is less responsive to network congestion.

An automated detection of service degradation involves simply comparing the client-prescriptions of expected QoS and the parameters of actual QoS offering from the network infrastructure [5]. A monitor embodies a set of domain-specific procedures that are activated through a meta-level signaling from management stations. These procedures sample the service interface state, and then evaluate a system-wide predicate to detect the occurrence of QoS changes. Such a computational detection of QoS events can be realized over a much faster time-scale. This allows a timely recovery from, say, infrastructure resource failures.

When the client application actions to deal with QoS degradations are expressible in a closed-form (using domain-specific procedures), the client and service agents coordinate with each other to reconfigure the application. These functions depict the *policies* on how to react to the changes occurring in various operating regions of the network service. Here, application adaptation refers to fine-granular changes in QoS specs (say, by adjusting the parameter values) over a small linear operating region. Whereas, application reconfiguration refers to coarse granular changes, at the policy function level, arising due to wider swings into non-linear operating regions. A repertoire of policy functions may prescribe reconfigurations in terms of different QoS parameters and/or logical compositions of QoS specs.

2.4   Reconfiguration Management Module

The management module (RSMM) maps client-initiated service requests onto
a specific protocol to control the network infrastructure. For this purpose, the
RSMM maintains the binding information for various services provided by the
infrastructure. Our focus is on the RSMM functionalities: non-intrusive service
monitoring and coordination of client adaptation, to support dynamic settings
where changes/outages in infrastructure resources may occur at various points
in time. If the RSMM does not provide service monitoring, detection of service
degradation by clients is possible only over problem-specific time-scales. On the
other hand, RSMM support for service monitoring enables client adaptations to
occur at computational time-scales.

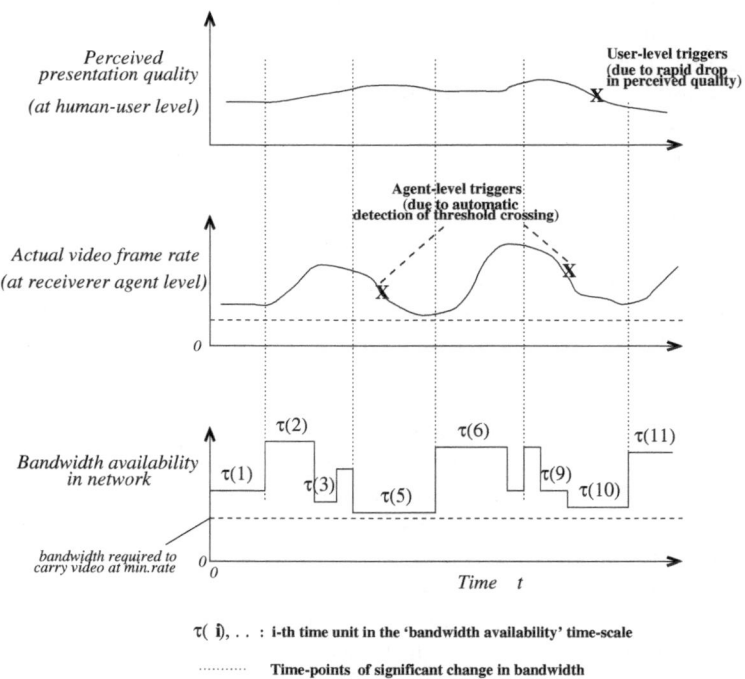

Fig. 2. Time-scales of parameter variations in a video distribution network

The service monitoring and client coordination roles of RSMM should be independent of the problem-domain an application may pertain to — so that they
can be deployed across diverse applications. For this purpose, the RSMM instantiates service-level meta-data dissemination algorithms with the application-
specific parameters (i.e., client-prescribed service attributes and their mappings
to infrastructure resource control protocol). The time-scale requirements of monitoring and adaptation are captured in these specifications.

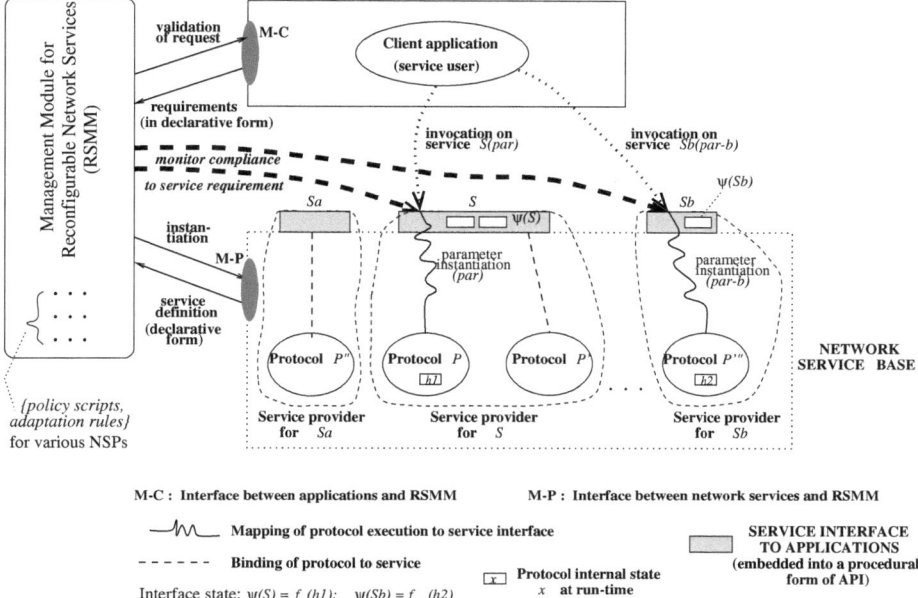

Fig. 3. Architectural view of service management system

Figure 3 shows an architectural view of how different services can be supported through a general-purpose management system. An application X is presented with a service interface $I(S)$ through which X prescribes a set of desired QoS parameters a. A protocol providing the service S, $P(S)$, is projected onto $I(S)$ in the form of a service state $\psi_X(S)$ — with a projection relation $f_P(h)$ mapping the protocol-internal state h onto the interface state, i.e., $\psi_X(S) = f_P(h)$. Any dynamic changes occurring in the operations of a NSP are seen by a management entity as *service-level events*, expressed as a predicate $\Phi(\psi_X(S))$. The RSMM basically monitors the service-level events to determine the course of recovery actions (through policy functions).

We now describe a distributed network application: rate-controlled video multicast over Internet, that can be managed through the prism of our RSMM structure.

3   Sample Application: 'Rate Control' Based Video Multicast

Consider a set of sources $\{V_a, V_b, V_c, \cdots\}$ multicasting video packets at the rates $\{\lambda_y\}_{y=a,b,c,\cdots}$ to a set of receivers $\{R_1, R_2, R_3, R_4, \cdots\}$ over a network path that implements 'best-effort' delivery of packets (say, video conferencing over Internet) [2]. See Figure 4. With no reservation of link bandwidths along the path, the video packet flow is vulnerable to link congestion that may arise due to extraneous factors: such as interfering cross-traffic and bandwidth depletion on a link (marked as 'X' in Figure 4). A congested link can drop many packets in a flow, which is sensed by receivers fed through this link in the form of sustained packet misses relative to an expected sequence of packet arrivals. Since a 'best-effort' network does not track the congestion status on its links, receiver agents report to source agents about the packet loss experienced, as a means of congestion notification — say, using IETF RTCP. The report from each receiver agent $AM(i)|_{i=1,2,\cdots}$ indicates the source-itemized loss rates experienced at R_i, based on checks of the sequence numbers assigned by each source agent $AM(y)|_{y=a,b,\cdots}$ for the packets from V_y. Based on these loss reports, one or more sources reduce their send rates in an effort to relieve congestion.

3.1 Protocol Aspects of Video Rate Control

The Internet does not allow the application to see the multicast path (let alone control the path set up). So, when congestion occurs in one or more path segments, sources listed in the receiver loss reports do not know whose traffic traverses path segments already congested due to cross-traffic, or, whether their traffic solely causes the congestion. Sources not listed in the loss report of an agent $AM(i)$ however know that their traffic traverse only non-congested path segments in reaching R_i. Thus, instead of having all the sources reduce their rates, the protocol *infers* a coarsely approximate knowledge about the topology from the loss reports of various receivers and then has only the sources whose traffic flows through the congested paths reduce their send rate. Since video traffic is often bursty, the rate adjustment by sources may also consider the statistical multiplexing effects arising from a shared use of paths by various traffic in a given source-receiver configuration.

Referring to Figure 4, the congestion along the upstream path segment from V_b causes the loss of packets from V_b at all the receivers; whereas, the congestion in the downstream path segment shared by R_3 and R_4 causes the packets of all the sources V_a, V_b, V_c to be lost at R_3 and R_4 but not at R_1 and R_2. So, $[R_3, R_4]$ report a high data loss from V_b and a moderate data loss from $[V_a, V_b]$; whereas, $[R_1, R_2]$ report a moderate data loss from V_b and no loss from $[V_a, V_c]$. Accordingly, V_b reduces its send rate substantially, while V_a and V_c moderately reduce their send rates (the latter may take into account statistical multiplexing effects as well).

The sources implement a rate adjustment rule: exponential reduction in send rate when the path is congested, and additive increase in send rate when the path is not congested. The rate reductions/additions are realized by service agents over faster time-scales (say, every 5 *sec*). Whereas, other types of changes in the send rates of geographically separated sources are stipulated by the application itself — which occurs over larger time-scales and wider spatial scales. A procedural realization of the bandwidth-to-QoS mapping thus involves iterative adjustments of send rates, over short time-scales, until the congestion just gets relieved.

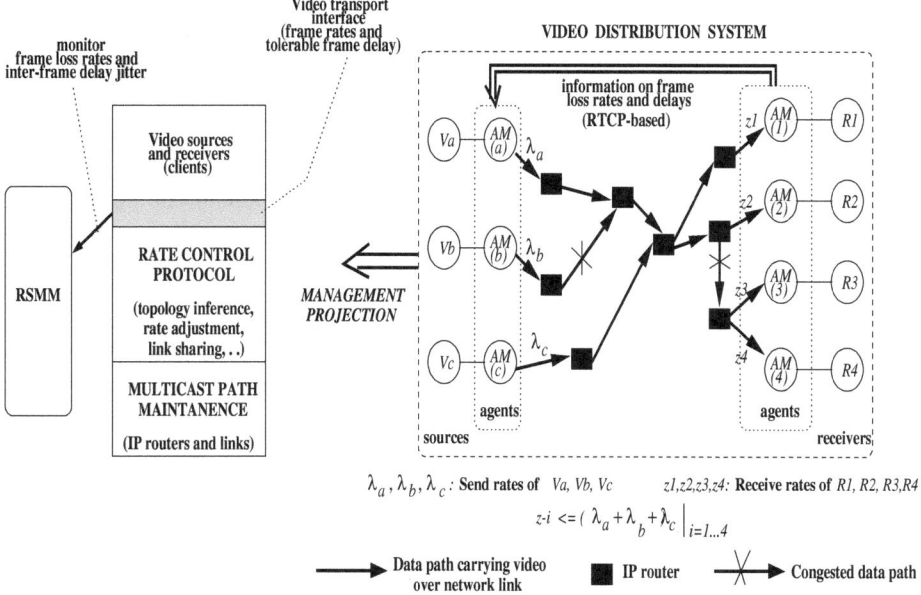

Fig. 4. Management view of video multicast

3.2   Management of Adaptive Video Multicast

For management purposes, the rate-controlled video multicast is seen as a network service VDIST. $\{V_a, V_b, \cdots\}$ and $\{R_1, R_2, \cdots\}$ form the client application, interacting with the agents through a service interface. The QoS specs are the video frame rates and loss tolerances $\{\lambda_y, \delta_y\}_{y=a,b,\ldots}$. The agents $\{AM(i), AM(y)\}_{\forall i,y}$ collectively implement the intra-NSP(VDIST) functionality of loss monitoring and source rate adaptation, on top of a multicast path set up by the native IP routing mechanism. In addition, the agents $\{AM(y)\}_{\forall y}$ implement also the RSMM functionality of application-level reconfigurations — such as source-receiver drop outs and differential rate adjustments of sources. Refer to Figure 4.

Based on the loss reports of receivers and where the offending sources are placed in the (inferred) topology relative to receivers, the agents $AM(y)|_{y=a,b,\cdots}$ compute a schedule for adjusting the send rates using the AIMD ('additive increase multiplicative decrease') algorithm:

$$\lambda_y'(j+1) = \max(\{[\lambda_y'(j).exp(-\alpha_y \overline{L_y(j)})], \lambda_y^{(min)}\})$$
$$\text{for } \overline{L_y(j)} > \delta_y$$
$$\lambda_y'(j+1) = \min(\{[\lambda_y'(j) + \beta_y], \lambda_y^{(max)}\}) \quad \text{for } \overline{L_y(j)} < L_y^{(min)}, \tag{1}$$

where $j = 0, 1, 2, \cdots$ is the iteration number in a run-time execution of the rate adjustment algorithm (with $\lambda_y'(0) = \lambda_y$), $\overline{L_y(j)}$ is the average loss rate of data from source y observed across all receivers in j^{th} iteration $(y = a, b, \cdots)$, α_y and β_y are positive constants, and $L_y^{(min)}$ is a loss threshold such that $L_y^{(min)} < \delta_y$. The constants $(\alpha, \beta, L^{(min)})$ impact the adaptation performance such as stability, steady-state error, and convergence. The above rule, subject to rate 'min-max' limits, is itself loaded into the NSP(VDIST) as an 'applicative' function.

Policy scripts for rules with different (parameter,value) sets may contain the relevant performance annotations: such as a small α increases the convergence time, a large δ lowers the average presentation quality, and a small $(\delta - L^{(min)})$ increases the presentation jitter. RSMM may use these annotations to order/index the VDIST policy functions. Furthermore, alternate rules may consider other factors that are also pertinent to the video distribution: such as receiver priorities in computing $\overline{L}$, minimum guaranteed rate across all receivers, and quantified statistical multiplexing gains. The rules may be formulated based on, say, how fast and smooth the human users in application can react to the changes in video send rates. The rules are, in part, designed to weigh against stochastic variations in the cross-traffic that flows along the network paths carrying video — such as fluctuations in traffic intensity and time-scales.

3.3   Experimental Study of Video Multicast

The adjustment of video send rates of multicast sources is based on whether the congestion detected along a path segment is caused by the sources themselves, or, by an unrelated cross-traffic sharing the path. In the former case, a short-term adjustment of the send rates of sources is likely to relieve the congestion. Whereas, a cross-traffic induced congestion necessitates application-level reconfigurations such as having the affected sources and/or receivers leave the multicast session or lower their rate specs substantially.

The rate adaptation protocol de-correlates the two causes of congestion from the loss reports of receivers by first resorting to the short-term measure of reducing the send rates of all sources (possibly, with some priority rules) and then examining its impact on the inferred topology to decide on whether a reconfiguration is needed. Referring to Figure 4, suppose an across-the-board reduction of send rates (possibly, with a higher reduction from V_b) relieves congestion only in the path segment leading to R_3 and R_4. With an inference that congestion in the path segment leading from V_b is more likely due to cross-traffic, the application may reconfigure by, say, having V_b, say, leave the conference session. On the other hand, if across-the-board rate reduction relieves congestion only in the path leading from V_b, R_3 and R_4 may either leave the conference session or switch to a voice-only mode. A session reconfiguration thus requires application-level knowledge, and is resorted to only if the rate reduction initiated by the source and receiver agents does not succeed.

Figure 5 illustrates how the video packet flow rate, as seen by the sources and receivers, changes with the available network bandwidth. This behavior, monitored by the source agents using receiver loss reports, is based on empirical data collected from our studies of video rate control mechanisms with 2 sources and 12 receivers interconnected by an IP multicast path The maximum and minimum send rates are set as 20 *fps* and 3 *fps* respectively (with some form of linear mapping assumed between the send rates and the perceptual quality at human users). For simplicity, the average loss rate of a source experienced among the receivers is used in adjusting the send rate of this source. The reduction in the send rate of V_a is higher than that of V_b: because the path from V_a to the receivers has more congested segments, as inferred from the loss reports, than that from V_b.

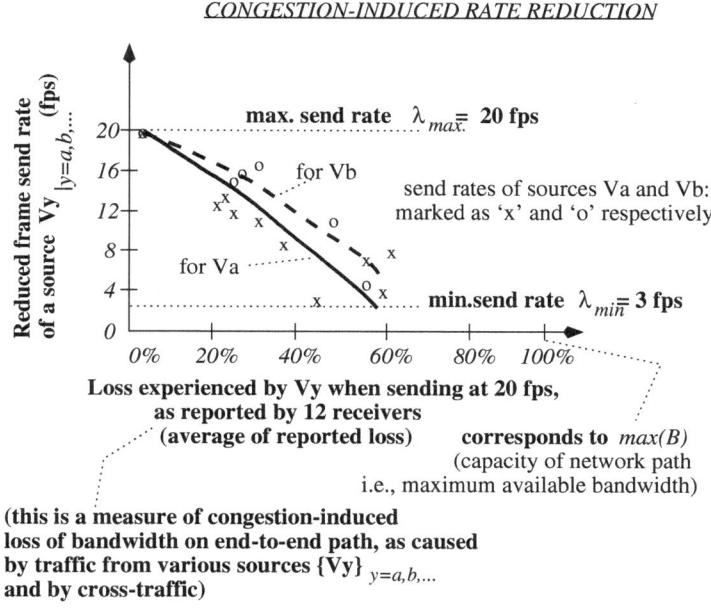

Fig. 5. Video send rate behavior under congestion

Our interest is in the monotonic convex behavior of the send rate with respect to congestion in the network. The latter is measured by how much of the available bandwidth B_{av} is taken by the video packets of various sources and the cross-traffic. The behavior data takes into account the statistical sharing of network bandwidth that (implicitly) occurs between various data traffic. The behavior data is useful for the rate adaptation algorithms in determining how fast and by how much a reduction in the send rates should occur.

As can be seen, the two steps involved in responding to congestion, namely, source rate reduction and session reconfiguration, occur over different time-scales, and require different levels of management intelligence.

3.4 Intelligent Control in Video Multicast

The physical world consists of the source and receiver agents carrying out the rate adaptation procedures as per the rule loaded into the application agents directly inter-working with the NSP. The procedures for estimating the available bandwidth of a multicast path are embodied in the NSP(VDIST). The distribution of rate reduction among the sources is determined by the short-term inferred congestion topology from receiver loss reports and the stable available bandwidth as estimated by NSP(VDIST) — c.f. Equation (1). The rate reduction achieved therein over multiple iterations of an adaptation cycle is treated by the RSMM as a single time-point execution of the NSP(VDIST).

It has been shown by Internet Measurements researchers that a cross-traffic often exhibits long-term stationarity [6]. Accordingly, the topology segments inferred as a result of cross-traffic induced congestion are stable over long time intervals — in comparison with that induced by the video traffic in the application on hand. The latter type of congestion is quite likely to be relieved by an across-the-board rate reduction by various sources as per the adaptation rule currently enforced by the source and receiver agents. Whereas, a cross-traffic induced congestion is very likely by application-level reconfigurations (such as the drop out of selected sources and receivers from the multicast session). The resulting QoS degradation effected across the various sources and receivers needs to be assessed in terms of the reduction in utility and its impact on the video multicast application. The management entity selects a suitable policy script and then loads it into the agents directly working the application to reconfigure the sources and receivers, thereby providing a context for the rate adaptation procedures executed at the lower level.

Figure 6 illustrates the partitioning of management intelligence among the 'video multicast' application components. The split of intelligence between the components is based on the scope and context of recovery actions: namely, at source rate control level or at source-receiver configuration level[1].

As can be seen, we infuse management intelligence in the control of adaptive network applications. Our function-based generic model of NSPs is useful for this purpose.

4   Existing Management Paradigms

There has been a number of works on policy-driven network management, with specific focus on adaptive QoS support. For instance, [7] describes dynamic policy mechanisms for characterization of network resources (such as link bandwidth and end-station CPU load). How these mechanisms can be extended to cover a broader class of virtualized resources in a context of service-level management is not clear. Likewise, [8] describes a resource query system for network-aware applications. However, this work also centers more on managing only the basic types of network resources.

---

[1] The split intelligence in video multicast applications is an instance of the *cyber-physical systems* paradigm that structures an embedded system as intelligent physical and computational worlds. See National Science Foundation announcement NSF 08-611, Oct. 2007, for a generalized description of 'cyber-physical systems'.

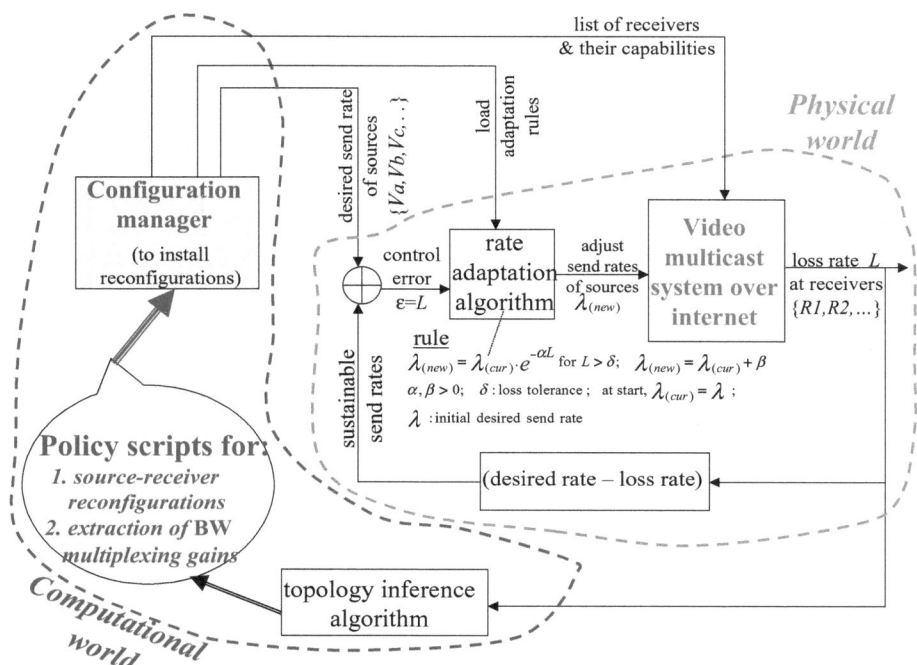

Fig. 6. Management intelligence in adaptive video multicast system

The work in [9] makes a case for autonomic network management based on an introspective view of the historical successes and failures faced by the designers of TCP congestion control and BGP routing protocols. This work outlines good principles for autonomic network designs, such as the explicit statement of environment assumptions and service-level goals and the need for built-in self-test procedures. The work however does not identify any concrete model or software engineering principles for autonomic management.

The SNMP-based model defines a set of APIs that can be invoked at a network management station, and also the underlying (signaling) message exchanges between the management station and the agents executing at the target network sites [10]. SNMP allows a prescription of packet filters based on 'logical expressions' that select which packets are captured for further analysis. An event table prescribes when a notification is to be sent to the management station. The packet selection criteria and the functions necessary to prescribe events therefrom are oriented towards traffic and fault analysis at various network elements (e.g., the ICMP packet loss rate exceeding a limit may indicate a path failure).

In contrast, the goal of our approach is to manage network services — rather than networks themselves (as advocated in [11]). Our approach is towards behavior analysis of network subsystems as seen by applications through service interfaces (and hence, is at a higher level than existing works). Our earlier work on service-oriented network management provides meta-level abstractions to develop the programming and signaling support for network services [12]. Our other earlier works [13,14] describe protocol-level adaptations, by parameter adjustments and/or by protocol switching, to offer network services with optimal performance and sustained availability. We extend our prior works by employing an *intelligent agent-based architecture* to manage the service provisioning under a wide range of application scenarios.

The management intelligence, as embodied in the RSMM, manifests partly in the application layer and partly in the network service layer. In an underlying implementation, the RSMM maintains policy scripts and rules for application-wide reconfigurations and short-term adaptations, prescribed over different QoS spaces and environment conditions. In this light, existing works on policy-based network management allow policy specification statically [15] or dynamically [16,17]. But they do not offer an unified intelligence substrate for policy selection based on a holistic analysis of the dynamically changing application needs, infrastructure resource availability, and external environment conditions.

Overall, our approach of split management intelligence offers a new framework to design reconfigurable network services.

5   Conclusions

When developing a diverse range distributed network services, challenges arise due to sub-system level failures and/or changes that may dynamically occur and can, in turn, affect the application functionality. Here, a major challenge is posed by the need to automatically evaluate the service behaviors, in terms of QoS, during various stages of service provisioning to client applications.

To support the integration of network service management tools into distributed application development environments, we employed a paradigm founded on modular decomposition principles. In this paradigm, a service may be provided by a protocol module through a generic interface, with the client application instantiating this module with a desired set of parameters. The reconfiguration management module (RSMM) maintains the binding information for network services that can be provided through infrastructure resource usage. The RSMM supports a highly dynamic setting, such as changes/outages in service provisioning (i.e., QoS fluctuations) and application reconfigurations that may occur over different time-scales. The RSMM effectively 'brokers' between the applications and network services to coordinate their interactions in a flexible and configurable manner.

Our paper described a new model of service provisioning, using principles of 'function'-based decomposition of service components across well-defined interfaces. The 'function'-based model separates a service interface to clients from the details of protocol modules that actually provide the service. Critical properties may be associated with a service interface, prescribable in the form of composition relations on service attributes. Client invocations on a service may instantiate these attributes with parameters, along with a prescription of what service property should hold. The RSMM incorporates a monitoring functionality that checks for compliance to prescribed critical properties at run-time. The RSMM maintains application-supplied policy scripts and rules to coordinate the reconfigurations and adaptations needed, should a service property violation occur at run-time (due to dynamically occurring changes in the infrastructure resources and/or the external environment). The paper also described the case study of an application: rate-adaptive video multicast distribution over IP networks.

Overall, our service model takes into account the cross-layer interactions between the network infrastructure, distributed protocols that adaptively exercise the network, and application-level reconfigurations in response to network conditions as seen through a service interface. To enable this holistic approach to application designs, our management entity, RSMM, brokers a quantifiable means of access to network services (using a repository of application-supplied policies and rules). The splitting of management roles between the service and application layers offers scalability and flexibility of our network service model, while reducing the costs of distributed network software development.

References

1. Chen, Y., Katz, R.H., Kubiatowicz, J.: Dynamic replica placement for scalable content delivery. In: Druschel, P., Kaashoek, M.F., Rowstron, A. (eds.) IPTPS 2002. LNCS, vol. 2429, p. 306. Springer, Heidelberg (2002)
2. Diot, C., Huitema, C., Turletti, T.: Multimedia Applications Should be Adaptive. In: Proc. HPCS 1995, Mystic (CN) (August 1995)
3. Ravindran, K., Steinmetz, R.: Object-oriented Communication Structures for Multimedia Data Transport. IEEE Journal on Selected Areas in Communications 14(7), 1360–1375 (1996)
4. Grant, T.: Unifying Planning and Control using an OODA-based Architecture. In: proc. SAICSIT (2005)
5. Giallonardo, E., Zimeo, E.: More Semantics in QoS Matching. In: Proc. IEEE Intl. conf. on Service-oriented Computing and Applications (SOCA 2007) (2007)
6. Zhang, Y., Duffield, N., Paxon, V., Shenker, S.: On the Constancy of Internet Path Properties. In: Proc. SIGCOMM Internet Measurement Workshop (2001)
7. Howard, S., Lutfiyya, H., Katachabaw, M., Bauer, M.: Supporting Dynamic Policy Change Using CORBA System Management Facilities. In: Proc. Integrated Network Management, IM 1997, San Diego, CA (May 1997)
8. Gross, T., Steenkiste, P.: J. Subhlok. Adaptive Distributed Applications on Heterogeneous Networks. In: Proc. 8th Heterogeneous Computing Workshop, HCW 1999 (1999)
9. Mortier, R., Kiciman, E.: Autonomic Network Management: Some Pragmatic Considerations. In: SIGCOMM 2006 Workshops, Pisa (Italy), ACM, New York (2006)
10. Subramanian, M.: SNMP Management RMON. In: Network Management: Principles and Practice, ch. 8. Addison-Wesley Publ., Reading (2000)
11. Erfani, S., Lawrence, V.B., Malek, M.: The Management Paradigm Shift: Challenges from Element Management to Service Management. Bell Labs Technical Journal 5(3), 3–20 (2000)
12. Ravindran, K., Liu, X.: Service-level management of adaptive distributed network applications. In: Malek, M., Reitenspiess, M., Kaiser, J. (eds.) ISAS 2004. LNCS, vol. 3335, pp. 101–117. Springer, Heidelberg (2005)

13. Ravindran, K., Wu, J.: Architectures for Protocol-level Adaptations for Enhanced Performance of Distributed Network Services. In: Proc. Network Operations and Management Symposium, NOMS 2006. IEEE-IFIP, Los Alamitos (2006)
14. Ravindran, K.: Dynamic Protocol-level Adaptations for Performance and Availability of Distributed Network Services. In: Modeling Autonomic Communication Environments. Multicon Lecture Notes (October 2007)
15. Verma, D., Beigi, M., Jennings, R.: Policy-based SLA Management in Enterprise Networks. In: Res. Report, IBM T. J. Watson Res. Center (2001)
16. Lymberopoulos, L., Lupu, E., Sloman, M.: An Adaptive Policy-based Framework for Network Services Management. Journal of Network and Systems Management 11(3) (September 2003)
17. Chhetri, M., Kowalczyk, R.: Agent Enabled Adaptive Management of QoS Assured Provision of Composite Services. Cybernetics and Systems: an Intl. Journal, Special Issue on Applied Intelligent Systems (2008)

Business Modeling via Commitments

Pankaj R. Telang and Munindar P. Singh

Department of Computer Science
North Carolina State University
Raleigh, NC 27695-8206, USA
prtelang@ncsu.edu, singh@ncsu.edu

Abstract. Existing computer science approaches to business modeling offer low-level abstractions such as data and control flows, which fail to capture the business intent underlying the interactions that are central to real-life business models. In contrast, existing management science approaches are high-level but not only are these semiformal, they are also focused exclusively on managerial concerns such as valuations and profitability.

This paper proposes a novel business metamodel based on commitments that considers additional agent-oriented concepts, specifically, goals and tasks. It proposes a set of business patterns and algorithms for checking model completeness and verification of agent interactions. Unlike traditional models, our approach marries rigor and flexibility, providing a crisp notion of correctness and compliance independent of specific executions.

1 Introduction

Real-life service engagements generally involve long-lived, complex interactions among two or more autonomous business partners. We define a *business model* as a specification of a way in which a service engagement is carried out. We address the problem of creating, enacting, and verifying business models from a high-level, yet rigorous standpoint.

Service organizations form complex business relationships with other organizations to exchange value. Competition continually forces organizations to improve their operations. Such improvements include outsourcing or insourcing business tasks based on appropriate strategic considerations. Mergers, acquisitions, and alliances change the partners of a value network. The business processes needed to support such dynamic interactions tend to be complex.

Existing techniques for modeling, operationalizing, and evolving such processes are inadequate, because they are based on low-level abstractions at the level of data and control flows, expressed in orchestrations or choreographies. These specifications do not capture the business intent of the interactions. They tend to over-constrain business behavior by mandating the exchange of a predetermined set of messages usually in an unnecessarily restrictive temporal order.

This paper proposes a commitment-based business metamodel, which captures value exchanges among business partners in terms of their commitments. Further, this paper defines patterns based on the above metamodel as well as algorithms to verify the correctness of service engagements with respect to their designs. The main benefits of

R. Kowalczyk et al. (Eds.): SOCASE 2009, LNCS 5907, pp. 111–125, 2009.

this approach are to meld rigor and flexibility thereby improving the quality of service engagement specifications and their instantiations.

Contributions. The main contributions of this paper are (1) a commitment-based meta-model that describes value exchanges among business partners, (2) a set of business modeling patterns, and (3) algorithms for verifying (a) implemented agent interactions with respect to a business model and (b) the completeness of a business model.

Organization. Section 2 presents the business metamodel and a set of business patterns. Section 3 applies the patterns to create a model for an insurance claim processing scenario. Sections 4 and 5 introduce notions of compliance and completeness respectively, and provide algorithms for checking them. Section 6 compares our approach with related work.

2   Metamodel and Patterns

A business model seeks to capture value exchanges and the evolution of commitments among business partners. Figure 1 illustrates our metamodel.

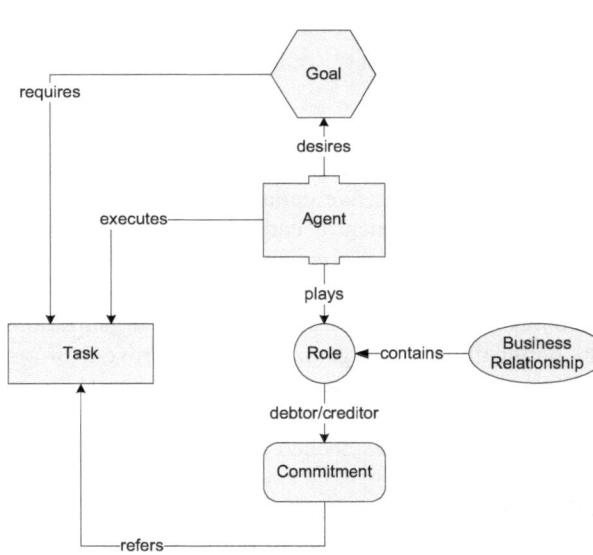

Fig. 1. Metamodel for commitment-based business models

We characterize a business model via a set of *business relationships*, the participants of which we term *(business) partners*. The partners execute tasks for each other that enable achieving their respective *goals*. Importantly, our approach defines relationships in terms of the creation and manipulation of *commitments* among the partners. To enter into a business relationship, each partner takes on the commitments that the relationship specifies. The partner presumably possesses the *capabilities* that the relationship requires—these are presumably required to perform the *tasks* that would discharge the specified commitments.

We associate interaction protocols with business relationships in two main ways. Interaction protocols are crucial both to (1) creating or modifying a business relationship, such as via negotiation and (2) to enacting a business relationship. The following paragraphs describe the key concepts of this metamodel.

Agent: a computational representation of a business partner. An agent captures the autonomy and heterogeneity of a real-world business. An agent has goals and possesses a set of capabilities that enable it to execute business tasks. For each business relationship in which an agent participates, it enacts one or more roles in that relationship.

Role: an abstraction over agents that helps specify a business relationship. Each role specifies the commitments expected of the agents who play that role along with the capabilities they must possess to function in that role.

Goal: a state of the world that an agent desires to be brought about [3]. In simple terms, an agent's goals are its ends. An agent achieves a goal by executing appropriate tasks.

Task: a business activity viewed from the perspective of an agent. Value transfers between the agents when they execute tasks for one another.

Capability: an abstraction of the tasks that an agent can perform.

Commitment: A commitment C(DEBTOR, CREDITOR, antecedent, consequent) denotes that the DEBTOR is obliged to the CREDITOR for bringing about consequent if antecedent holds [8]. A commitment C(BUYER, SELLER, goods, pay) means that buyer commits to paying the seller if goods are delivered. When the seller delivers goods, the buyer becomes unconditionally committed to paying. In the event that the buyer makes the specified payment, this commitment is discharged.

Business relationship: a set of interrelated commitments among two or more roles that describe the value to be exchanged among the roles. In simple terms, each agent's main motivation behind forming a business relationship is to access the capabilities of others.

At run-time, commitments arise between agents, but at design-time we specify them between roles. Being able to manipulate commitments yields the flexibility needed in open interactions. A commitment may be *created*. When its consequent is brought about, regardless of whether antecedent holds or not, it is *discharged*, i.e., satisfied. If its antecedent is brought about then it is *detached*. The creditor may *assign* a commitment to another agent. Conversely, a debtor may *delegate* a commitment to another agent. A debtor may also *cancel* a commitment and a creditor may *release* the debtor from the commitment. Further, a commitment moves among four main states: *active* (when it is created and (presumably) being worked upon), *pending* (when it has been delegated and is not being worked upon), *satisfied*, and *violated*.

2.1 Running Example

We evaluate the proposed metamodel and patterns via a real-world insurance claim processing use case involving AGFIL, an insurance company in Ireland [4]. AGFIL underwrites automobile insurance policies. Figure 2 shows the parties and processes involved in the business service of (emergency) claim processing that AGFIL provides.

To provide this service, AGFIL must provide claim reception and vehicle repair to the policy holders. Additionally, it needs to assess claims to protect against fraud. AGFIL depends on its partners, Europ Assist (EA), Lee Consulting Services (CS), and repairers, for executing these tasks. EA provides a 24-hour helpline for customers to

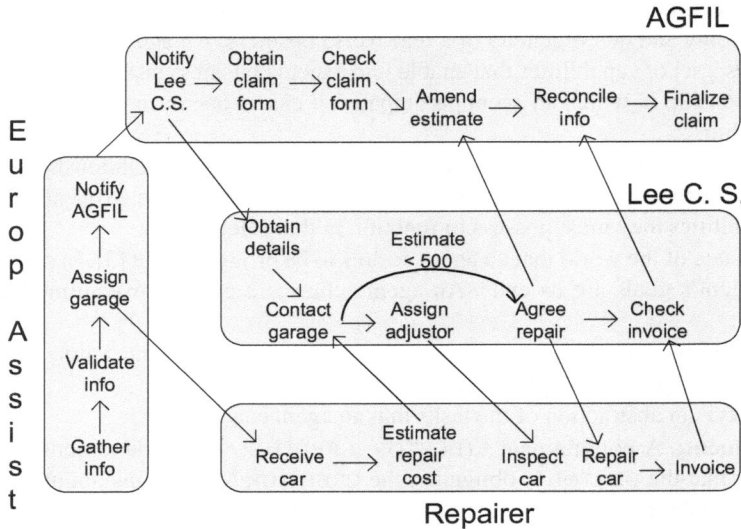

Fig. 2. Insurance claim processing [4]

report a claim and provides an approved repairer garage. CS assesses and presents invoices to AGFIL on behalf of the repairers. A network of approved repairers provide repair services. AGFIL retains the authority for issuing final claim approvals.

2.2   Patterns

A pattern, in the present setting, is a recipe for modeling recurring business scenarios. This section describes a key set of such patterns, which could seed a potential business model pattern library. Section 3 demonstrates the effectiveness of this simple set of patterns on an existing use-case based on a real-life scenario.

Of the 13 attributes in the classical template for object-oriented design patterns [6], we use *name*, *intent*, *motivation*, *implementation*, and *consequences* to describe our patterns. Here the *consequences* of a pattern allude to the practical consequences of applying the pattern, i.e., the assumptions underlying the model. The pattern figures use the notation of Fig. 1, and additionally show two directed edges for each commitment: from the debtor to the commitment and from the commitment to the creditor. The subscript on a commitment indicates its state: **A** for active, **D** for detached, **S** for satisfied, and **P** for pending. The patterns are expressed in terms of roles and would be instantiated by the agents who adopt the specified roles. Each role of a pattern must be adopted by some agent in order for the resulting business relationship to be executed.

2.3   Unilateral Commitment

Intent: A performer commits to a beneficiary for value transfer. There is no "converse" commitment from the beneficiary.

Motivation: For example, a conference committee member commits to a program chair to review a paper that the program chair asks the member to review. The chair makes no converse commitment.

Implementation: A commitment is created from the performer (R_1) to the beneficiary (R_2) for a value transfer. Figure 3 shows this pattern.

Consequences: This presumes a side benefit to the performer (debtor) from the antecedent of the commitment.

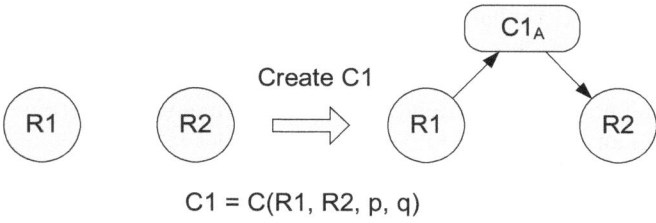

$$C1 = C(R1, R2, p, q)$$

Fig. 3. Unilateral commitment

2.4   Commercial Transaction

Intent: This pattern expresses a value exchange between two trading partners. The trading partners negotiate and, upon agreeing, commit to each other for the specified value transfers.

Motivation: A typical barter motivates this pattern. For example, a seller and a buyer agree to exchange goods for payment. A more conventional barter would be when the parties exchange goods and services rather than money for goods or services.

Implementation: A pair of reciprocal commitments between the trading partners (R_1 and R_2, treated symmetrically) specify the pattern. Figure 4 shows this pattern.

Consequences: In general, the antecedents and consequents of the commitments are both composite expressions. Importantly, we need a mechanism to ensure progress by in essence breaking the symmetry, e.g., via a form of concession [10].

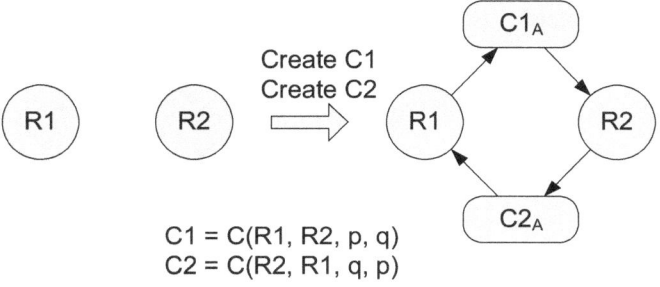

$$C1 = C(R1, R2, p, q)$$
$$C2 = C(R2, R1, q, p)$$

Fig. 4. Commercial transaction

2.5   Outsourcing

Intent: An outsourcer delegates a task to a subcontractor, typically because the outsourcer lacks the necessary capabilities or expects some other benefit such as a more efficient solution or a lower risk of failure.

Motivation: Many business organizations outsource noncore activities. As an example, consider a customer who signs up for cable television service. The cable operator commits to the customer for installation. Instead of staffing its entire service area directly, the cable operator outsources the installation task in several regions to its local partners in those regions.

Implementation: The outsourcer is the current debtor (R_1). The current debtor and the new debtor (R_2) create a relationship, following which the current debtor delegates the commitment to the new debtor. The existing commitment becomes pending; the new commitment becomes active. The creditor is unchanged. Figure 5 shows this pattern.

Consequences: The business relationship between the new and the previous debtors would be a standing arrangement, which must have a scope and lifetime no smaller than that of the delegated commitment. The commitment from the previous debtor is pending and must either be considered discharged or reactivated depending on how the new debtor performs.

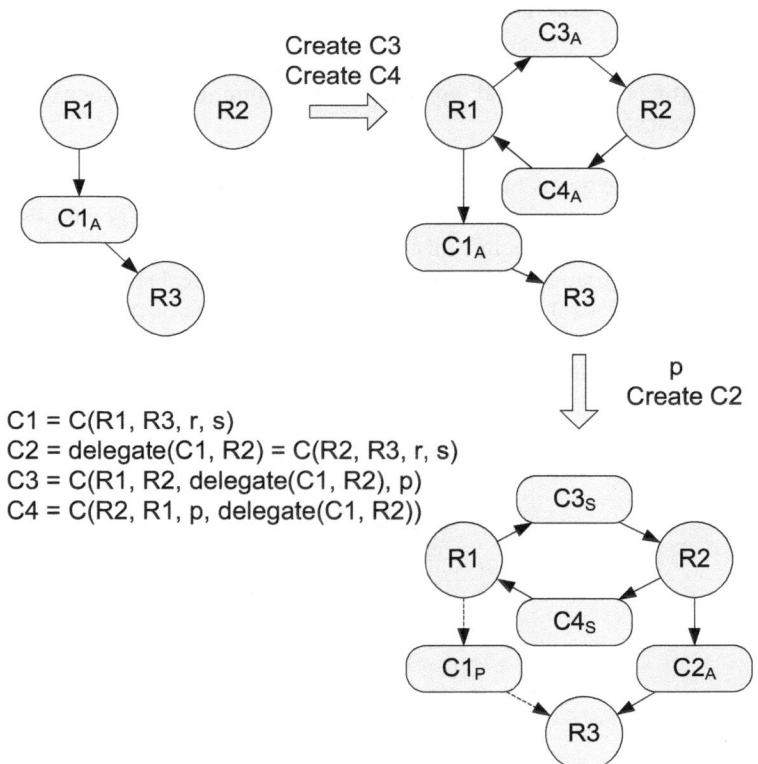

C1 = C(R1, R3, r, s)
C2 = delegate(C1, R2) = C(R2, R3, r, s)
C3 = C(R1, R2, delegate(C1, R2), p)
C4 = C(R2, R1, p, delegate(C1, R2))

Fig. 5. Outsourcing

2.6   Standing Service Contract

Intent: A service provider negotiates with a consumer for providing service over a specified duration, and creates a pair of commitments. The consumer's request for a service instance detaches the standing commitment. The provider then creates one or more commitments for providing the service instance.

Motivation: A business service such as plumbing maintenance or a line of credit from a bank refers to (potentially) numerous service instances. Whenever the faucet leaks (within specified limitations), the plumber will fix it. Whenever the customer submits a check for an amount up to the specified credit limit, the bank will disburse funds.

Implementation: The service provider (R_1) and consumer (R_2) enter into the following commitments. Here, C_1 and C_2 are reciprocal commitments (as in the commercial transactions pattern) that describe the standing service contract. C_3 and C_4 arise from the consumer exercising the service contract. Figure 6 shows this pattern.

Consequences: The standing contract must be of sufficiently large scope to cover the cases of interest but should generally be bounded in the effort it requires. This pattern can be applied multiple times as when a consumer pays a subscription every month to obtain a continuing plumbing warranty.

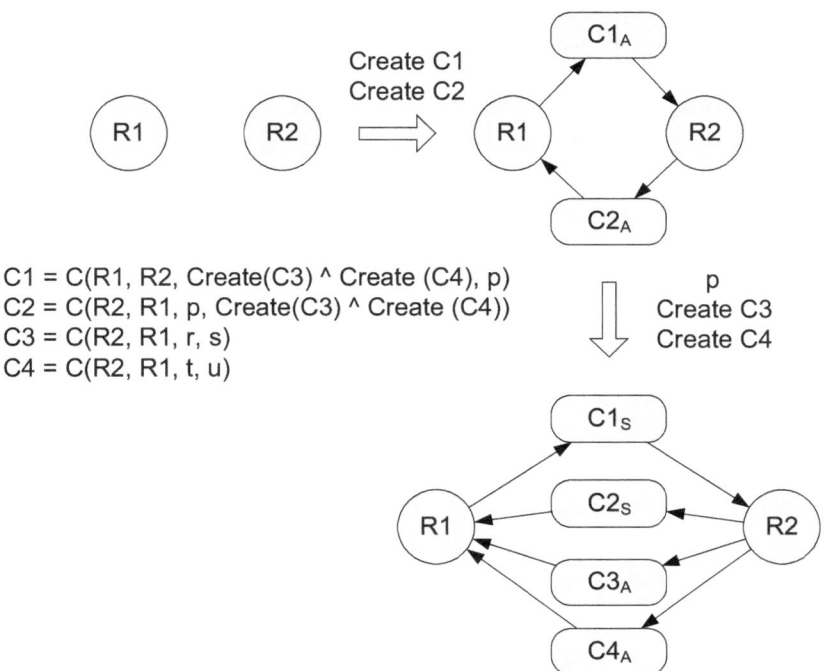

Fig. 6. Standing service contract

3   AGFIL Business Model

This section applies the patterns to the AGFIL scenario and describes the resulting business model. AGFIL, an insurer (I), has the goal of providing emergency service, which requires the capabilities for claim reception, claim assessment, claim finalization, and vehicle repair. Except claim finalization, which it possesses locally, AGFIL acquires the remaining capabilities from its business partners.

The insurer delegates to the call center its claim reception commitment to the policy holder. Although the commitment from the insurer to the policy holder for claim reception is not created yet, the insurer chooses to set up the delegation earlier. The *outsourcing pattern* models this scenario. The insurer selects EA as a call center provider (C). The selection process is out of our present scope. Figure 7 shows how the outsourcing pattern applies.

> C1. C(C, I, payCallcenter, create(C3))
> C2. C(I, C, create(C3), payCallcenter)
> C3. C(C, P, reportAccident, receiveClaim)

The insurer and the call center create commitments C1 and C2 when they agree upon the payment that the insurer makes to the call center, for providing claim reception to the policy holder. The commitment C1 means the call center commits to the insurer for creating commitment C3, which is to receive claims from the policy holder, provided the insurer pays the call center. The commitment C2 means the insurer commits to the call center for payment if the call center creates C3. The insurer pays the call center, and therefore discharges C2 and detaches C1. Later, the call center creates C3 and discharges C1.

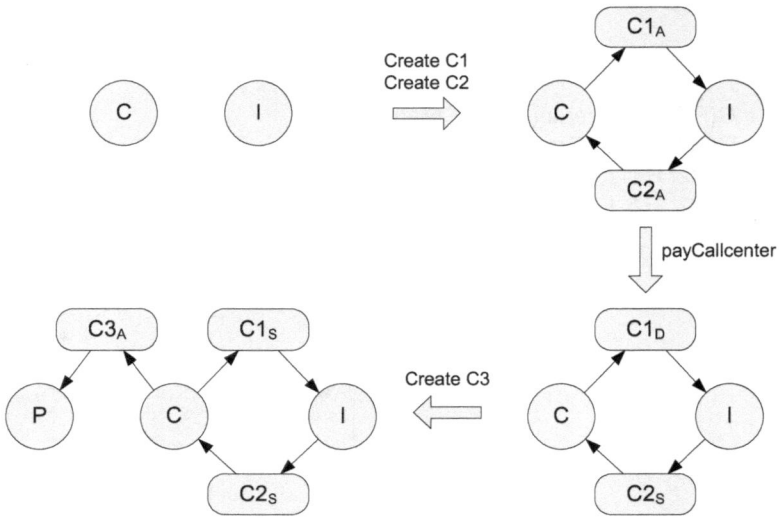

Fig. 7. Claim reception: Outsourcing

The insurer outsources the claim assessment capability to Lee CS, an assessor. In this case, the *outsourcing pattern* does not apply since the insurer is not delegating a commitment. That is, the insurer requires claim assessment for itself. Instead, the *commercial transaction pattern* models this scenario.

C4. C(A, I, payAssessor ∧ reqAssessment, agreeToRepair)
C5. C(I, A, agreeToRepair, payAssessor)

The commitment C4 means the assessor commits to the insurer, for negotiating repair cost and to bring about the agreement to repair with the repairer, provided the insurer pays the assessor and makes a request for assessment. The commitment C5 means the insurer commits to the assessor for the payment provided the assessor brings about agreement to repair.

The assessor outsources the vehicle inspection to an adjuster (D). The *commercial transaction pattern* models this scenario. Since this scenario is similar to the claim assessment scenario, to save space, we do not describe it in detail.

A policy holder (P) desires to get insurance. Through a directory service, the policy holder locates AGFIL, the insurer. The policy holder and the insurer interact to setup the insurance service contract. The *service contract pattern* models this scenario. Figure 8 shows how the service contract pattern applies.

C8. C(P, I, insurance, payInsurer)
C9. C(I, P, payInsurer, insurance)
C10. C(I, P, reportAccident, receiveClaim)
C11. C(I, P, requestService, repairVehicle)

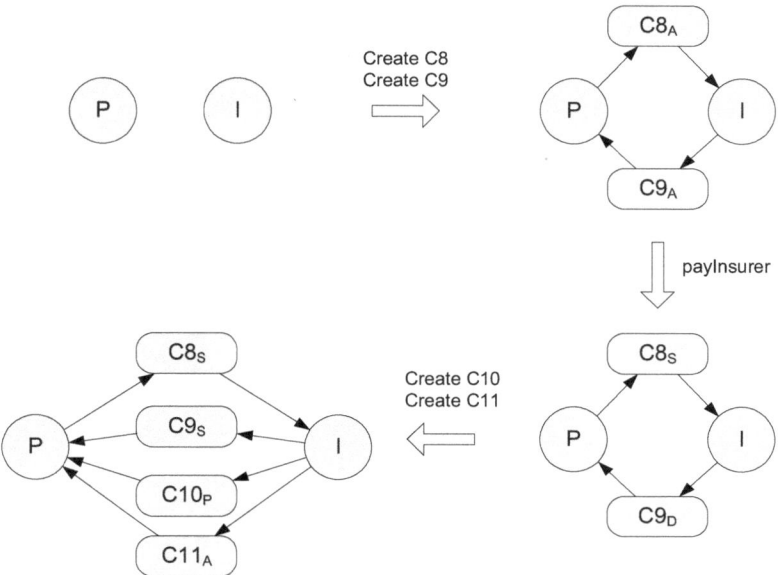

Fig. 8. Insurance purchase: Service contract

Commitment C8 means the policy holder commits to the insurer for payment if insurance is provided, and commitment C9 means the insurer commits to the policy holder for insurance if the policy holder pays the insurer. To provide insurance, the insurer creates the commitments C10 and C11, that is, insurance = create(C10) ∧ create(C11). Commitment C10 means the insurer commits to receiving claim if the policy holder reports an accident, and in commitment C11, the insurer commits to repairing the (insured) vehicle if the policy holder requests repair service for it. The insurer changes the status of commitment C10 to pending, since it has delegated that commitment to the call center. Recall that C3 results from the delegation of C10. That is, C3 = delegate(C10, C).

To assess a claim, the assessor has the adjuster inspect the vehicle. The assessor negotiates with the repairer. By bringing about an agreement to repair, the assessor satisfies its commitment to the insurer C4. Figure 9 shows how the *outsourcing pattern* now applies between the insurer, the repairer, and the policy holder.

C12. C(I, R, delegate(C11, R) ∧ agreeToRepair, payRepairer)
C13. C(R, I, payRepairer, delegate(C11, R))
C14. delegate(C11, R) = C(R, P, requestService, repairVehicle)

Commitment C12 means the insurer commits to the repairer for paying the repair charges, if the repairer accepts the delegation of C11 and creates C14. Commitment C13 means the repairer commits to accepting the delegation of commitment C11 if the insurer pays. In the delegated commitment C14, the repairer commits to the policy holder

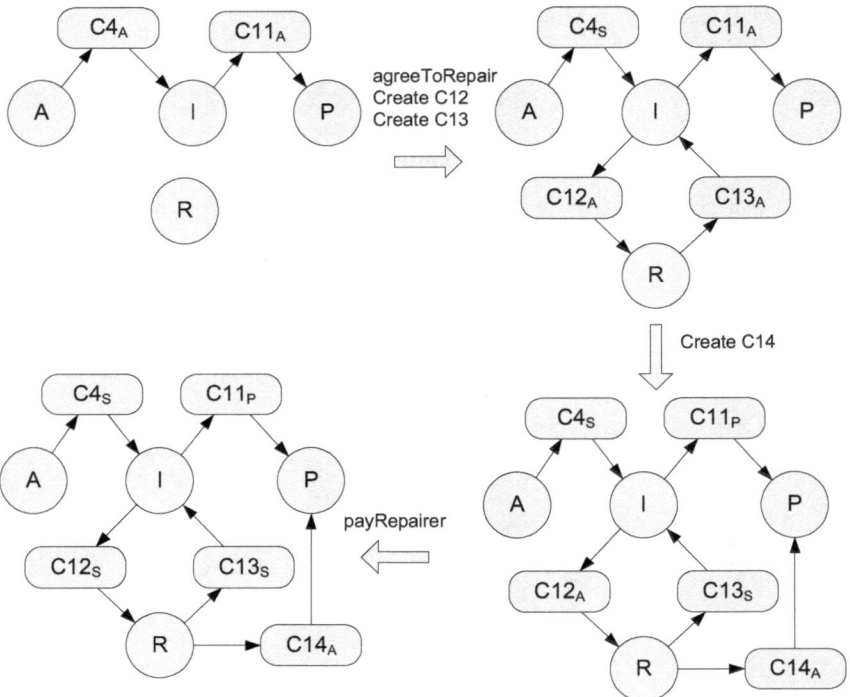

Fig. 9. Vehicle repair: Outsourcing

for vehicle repair when the policy holder requests for repair. The repairer satisfies the commitment C13 by creating C14, and detaches C12. Later the insurer discharges C12 by paying the repairer. Note that it is not necessary for the insurer to pay the repairer at this time, and other evolutions are possible. For example, the repairer may repair the vehicle, that is, satisfy the commitment C14, before the insurer pays. We describe one possible model evolution above.

4   Verifying Agent Interactions

This section presents an algorithm for verifying if each partner complies with a business model. An agent complies with a business model if it discharges each detached commitment of which it is the debtor. We consider a UML sequence diagram as a low-level model for agent interactions. The agents may exchange multiple messages for executing one task. For example, the policy holder may report an accident by sending a message to the insurer; the insurer may request additional information, leading to further messages. In the interaction model (based on a sequence diagram), we assume that upon completing a task, the executor of the task sends a message asserting its completion.

Algorithm 1. verifyInteractions(m, i): Verify agent interaction model i with respect to business model m

1 $C = m.C$; // Model Commitments
2 $CS = ()$; // Satisfied commitments
3 $CV = ()$; // Violated commitments
4 $T = i.T$; // Tasks completed in the interaction model
5 **foreach** $c \in C$ **do**
6   **if** *(eval(c.consequent, T) = true)* **then**
7          $CS.add(c)$;

8 **foreach** $((c \in C) \wedge (c \notin CS))$ **do**
9   **if** *(eval(c.antecedent, T) = true)* **then**
10          $CV.add(c)$

11 **return** CV;

Given a business model and an interaction model, Algorithm 1 returns a set of violated commitments. We assume that the interaction model captures all agent interactions. The algorithm iterates over the commitments from the business model and evaluates the antecedent and consequent of each using the tasks asserted in the interaction model. The antecedent and consequent of a commitment are formulae, each containing a disjunction of tasks. The $eval$ procedure evaluates these based on the tasks asserted in the interaction model. The commitments whose consequent evaluates to true are satisfied, whereas the commitments whose antecedent evaluates to true, but whose consequent evaluates to false, are detached commitments that are violated. The debtors of the violated commitments are the agents that do not comply with the given business model (within the scope of the given interaction model).

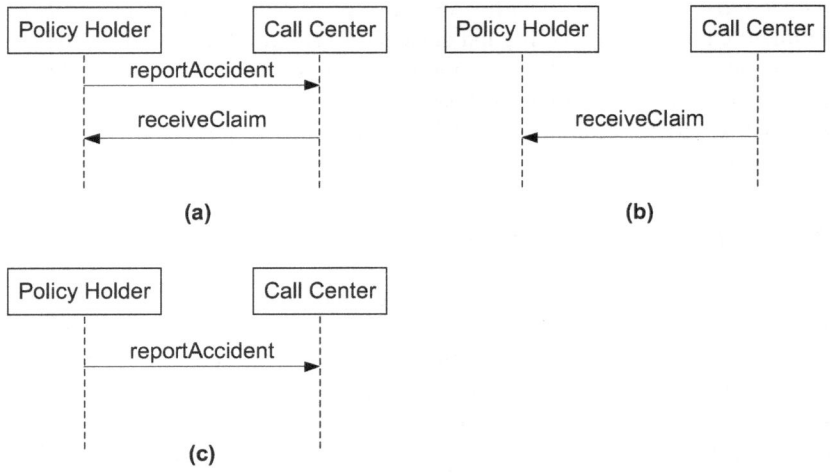

Fig. 10. Verifying agent interactions

For example, in the AGFIL business model, consider the commitment C10 = C(C, P, reportAccident, receiveClaim). An interaction model in which neither of the tasks, reportAccident and receiveClaim, are asserted, is a trivial case where both agents, the policy holder (P) and the call center (C), comply with the business model. In Fig. 10(a), the policy holder reports an accident, and detaches the commitment C10. The call center receives the claim and, therefore, satisfies the detached commitment C10. In this case, both the agents comply with the business model. In Fig. 10(b), the call center receives the claim and satisfies the commitment C10. This is another case where both the agents comply with the business model. In Fig. 10(c), the policy holder reports an accident, but the call center does not receive the claim. The call center violates the detached commitment C10, and lacks compliance with the business model.

5   Completeness

Agents enter into a business relationship for achieving their respective goals. A business model in which all agents achieve their goals is complete. It is important to check for model completeness, since in its absence, some agents will not achieve their goals and therefore desire to leave the relationship. That is, the business model will not be stable.

The Algorithm 2 checks a business model for completeness. For each agent, the algorithm checks if the agent can achieve all of its goals. An agent a can achieve a goal g, if it can execute all the tasks required for that goal. In case where the agent cannot execute all the tasks required for its goal, the model must contain commitments from other agents to execute the remaining tasks. Additionally, the agent a should be able to execute the tasks specified in the antecedents of those commitments. In the model, if there is an agent who cannot achieve a goal, then the algorithm returns false indicating that the model lacks completeness. Otherwise, the algorithm returns true.

Algorithm 2. verifyCompleteness(m): Verify completeness of business model m

---

1 $C = m.C$; // Model Commitments
2 $A = m.A$; // Agents
3 **foreach** $(a \in A)$ **do**
4      $G = a.G$; // Agent goals
5   **foreach** $(g \in G)$ **do**
6          $GT = g.T$; // Tasks for goal
7          $AT = a.T$; // Agent tasks
8          $task$: **foreach** $((t \in GT) \wedge (t \notin AT))$ **do**
9   **foreach** $(c \in C)$ **do**
10   **if** $((c.creditor = a)\wedge$
                   $(t \in tasks(c.consequent)) \wedge (tasks(c.antecedent) \subset AT))$ **then**
11   **next** $task$;
12   **return** false;

13 **return** true;

For example, consider the AGFIL business model. The assessor has the goal of claim assessment. To assess a claim, the assessor needs to inspectVehicle and agreeToRepair. The assessor has the capability of bringing about agreeToRepair, but it lacks the capability to inspectVehicle. In this case, for completeness, the model must contain commitment from some other agent to inspectVehicle. Additionally, the assessor should be able to bring about the antecedent of that commitment. For example, C(D, A, payAdjuster, inspectVehicle) is a commitment required for model completeness, assuming the assessor can perform payAdjuster.

6 Discussion

This section compares our approach with some existing approaches. Existing high-level approaches capture business organizations and value exchanges among them [1]. Many of these approaches are informal or semiformal and are developed for valuation and profitability analysis. They lack a rigorous treatment of business relationships (as via commitments) and lack a corresponding business-level notion of compliance.

Gordijn and Wieringa [7] propose the e^3-value approach, which captures a business organization as an actor. This is similar to the notion of an agent from our model. Actors execute value activities similar to the tasks in our model. In e^3, a value interface aggregates related in and out value ports of an actor to represent economic reciprocity. This concept is close to our concept of commitment, but it lacks formal semantics and doesn't yield equivalent flexibility. For example, unlike value interfaces, commitments can be delegated. Due to this, an e^3 model may capture value exchange among two actors, but during execution, the exchange and interaction may take place between two different actors.

Tropos [2] is an agent-oriented software methodology based on concepts of actor, goal, plan, and actor dependencies. The concepts of role, goal, and task from our model are similar to the Tropos concepts of actor, goal, and plan, respectively. A key difference between our model and Tropos is the concept of commitment. In Tropos, a dependency means that a depender actor depends on a dependee actor, for executing a plan or achieving a goal. This concept of dependency does not model what is required of the depender, and the dependee unconditionally adopts the dependency. Our debtor, creditor, and consequent are similar to the Tropos dependee, depender, and dependum, respectively. Unlike a dependency, a commitment includes an antecedent that brings it into full force. This allows modeling of reciprocal relationships between economic entities, which is lacking in the concept of dependency.

Opera is a framework for modeling multiagent societies [9], though from the perspective of a single designer or economic entity. In contrast, we model interactions among multiple entities. Opera's concepts of landmark, scene, and contract are close to our concepts of task, protocol, and commitment, respectively. However, Opera uses traditional obligations, which lack the flexibility of commitments.

Amoeba [5] is a process modeling methodology based on commitment protocols. This methodology creates model in terms of fine-grained messages and commitments. In contrast, our approach lies at a higher level of abstraction containing business goals, tasks, and commitments.

Conclusions. The main contributions of this paper are a business metamodel, a set of modeling patterns, and algorithms for verifying compliance and completeness of service engagements to business models. Our set of business model patterns is clearly not exhaustive; nor do we expect any set of patterns to be exhaustive—hundreds of patterns exist for programming and for software architecture, and the domain of business models is at least as complex as those. However, our core set of patterns shows how we may construct additional patterns. Future work includes development of a methodology for business modeling, model formalization and complexity analysis, and graphical tools for creating business models.

References

1. Andersson, B., Bergholtz, M., Edirisuriya, A., Ilayperuma, T., Johannesson, P., Gordijn, J., Grégoire, B., Schmitt, M., Dubois, E., Abels, S., Hahn, A., Wangler, B., Weigand, H.: Towards a reference ontology for business models. In: Embley, D.W., Olivé, A., Ram, S. (eds.) ER 2006. LNCS, vol. 4215, pp. 482–496. Springer, Heidelberg (2006)
2. Bresciani, P., Perini, A., Giorgini, P., Giunchiglia, F., Mylopoulos, J.: Tropos: An agent-oriented software development methodology. Autonomous Agents and Multi-Agent Systems 8(3), 203–236 (2004)
3. BRG. The business motivation model (2007)
4. Browne, S., Kellett, M.: Insurance (motor damage claims) scenario. In: Document Identifier D1.a, CrossFlow Consortium (1999)
5. Desai, N., Chopra, A.K., Singh, M.P.: Amoeba: A methodology for modeling and evolution of cross-organizational business processes. ACM Transactions on Software Engineering and Methodology, TOSEM (in press, 2009)
6. Gamma, E., Helm, R., Johnson, R., Vlissides, J.: Design Patterns: Elements of Reusable Object-Oriented Software. Professional Computing Series. Addison-Wesley, Reading (1995)

7. Gordijn, J., Wieringa, R.: A value-oriented approach to E-business process design. In: Eder, J., Missikoff, M. (eds.) CAiSE 2003. LNCS, vol. 2681, pp. 390–403. Springer, Heidelberg (2003)
8. Singh, M.P.: An ontology for commitments in multiagent systems: Toward a unification of normative concepts. Artificial Intelligence and Law 7, 97–113 (1999)
9. Weigand, H., Dignum, V., Meyer, J.-J.C., Dignum, F.: Specification by refinement and agreement: Designing agent interaction using landmarks and contracts. In: Petta, P., Tolksdorf, R., Zambonelli, F. (eds.) ESAW 2002. LNCS, vol. 2577, pp. 257–269. Springer, Heidelberg (2003)
10. Yolum, P., Singh, M.P.: Enacting protocols by commitment concession. In: Proceedings of the 6th International Joint Conference on Autonomous Agents and MultiAgent Systems (AAMAS), pp. 116–123 (May 2007)

MAMS Service Framework

Alexander Thiele, Silvan Kaiser, Thomas Konnerth, and Benjamin Hirsch

DAI Labor, Technische Universität Berlin
Ernst-Reuter-Platz 7
D-10587 Berlin
{alexander.thiele,silvan.kaiser,thomas.konnerth,
benjamin.hirsch}@dai-labor.de

Abstract. In this paper we describe the service execution platform for
the MAMS service framework which is an agent based platform for the
execution of services and service compositions. The MAMS framework
supports non-IT-experts in the process of generating, deploying and
executing new service compositions. It features an elaborate graphical
service creation environment, a service execution platform based on intelligent agents and it supports telecommunication specific functionalities
like the IP-Multimedia Subsystem. We have created a distributed service execution environment, which utilises agent technology to improve
scalability, platform management and stability. We will show how agent
technologies help attaining distributed service oriented systems.

1 Introduction

Nowadays, even though technologies for global provisioning of services are well
established, it is still difficult for small and medium enterprises (SME) to act as
service providers. The reasons for this come from the need for technical know-
how to create services and the necessary infrastructure to provide the created
services. The existing service authoring and creation tools are not intended to be
used by non-technical persons and the needed hardware as well as the software
to run the services must be purchased, configured and maintained, which makes
the provisioning of services costly if not unprofitable for small and medium
enterprises (SMEs).

The project MAMS (Multi Access - Modular Services) [16] addresses these
issues by allowing non-technical persons to quickly and easily create, deploy and
manage services, according to the users needs. Platform providers as technical
experts can concentrate on the provisioning of tools and basic services for service
creation, as well as on the infrastructure for reliable hosting of the services.

However, not only do the SME's need the means to create tailored services,
but the requirements for the service infrastructure in terms of flexibility and
configurability are very high too, as the service delivery platform needs to simultaneously provide for very different services running at the same time, with
different, often unpredictable loads. This is why the project followed an agent
based approach for designing the service delivery platform (ODSDP) that follows

R. Kowalczyk et al. (Eds.): SOCASE 2009, LNCS 5907, pp. 126–142, 2009.

tried and tested principles of service oriented architectures and implements them in an agent environment, thereby inheriting properties like flexibility, robustness, and scalability.

In this paper, we first give an overview about the state of the art in the area of service delivery platforms, easy service creation and agent technology. Before the new language for service composition (Section 4.2) and the agent-based platform for deployment, execution and management of services (Section 4) are described in detail, the project context of our solution will be explained in Section 3. Finally, experiences with the framework, a short conclusion and a precise outlook for the future work is given.

2   Related and Previous Work

Existing service delivery platforms, e.g. BEA WebLogic Communications Platform [3], IBM Service Provider Delivery Environment [12], Microsoft Connected Services Framework [19], or the HP Service Delivery Platform [11] do not fulfil all MAMS requirements [22]. In particular, the convergence between IT and telecommunication is not fully addressed. Platforms provide mobile services or focus on multi media services, depending on the economic background of the platform provider. HP introduces Web 2.0 and content management technologies in its service delivery technologies but separates them from the telecommunication services. They are also not sufficiently aware of the context and preferences of its users. New developments like semantic representation are not addressed. Another major drawback lies in the fact that it is not possible for non-technical persons to create services. Mostly, the provided graphical tools are too complicated and the platforms do not support the deployment of services by external service providers. Thus it is not possible for a non-IT-expert to create new services based on existing ones provided by the platform. In this sense the platforms are closed and a user is forced to take whatever the service provider has to offer. Though there is a trend to offer interfaces which enables external parties to provide additional content it does not overcome the main problem of being functionally enclosed.

There are also some European research projects in the area of easy service creation. OPUCE [21] provides a graphical tool based on a modelling language similar to UML for easy composition of services by non-professional service providers. SPICE [24] in addition considers semantical information to allow a more automatically composition of their rule-based services. LOMS [14] uses a template approach to support really non-technical service providers by integrating an additional service operator role. Revenue sharing between the different roles are considered by a multi-party business model. But these projects do not use agent technology to allow autonomous and flexible provisioning of services, which is in our view more appropriate to open, dynamic and unpredictable environments.

In the agent community, many researchers looked at the relevance of webservices to agents, for example [7,4,27]). The integration of webservices into agents has been looked at by e.g. Greenwood and Calisti [9] who developed a webservice gateway that mediates between FIPA agents and webservices. Zinnikus et al. [28] follow a similar approach where the agent's actions are exposed as webservices. Using these kind of approaches, a certain flexibility within service orchestrations can be achieved. However, service orchestration is more than the ability to use webservices, and the cited works where only of limited use for our project.

Others deal with the integration of workflows and agents, which is closer to the aims of the MAMS project. Already during the mid-90ties, the ADEPT project used agents that provided services and organised themselves to achieve business goals [1,20]. The focus here was however on the use of workflow and not so much on the technical issues like deployment, a vital element of MAMS.

Singh et al. used agents to manage workflows [23], and the WADE system [5] extends JADE to allow agents to orchestrate methods to workflows. The system can however not be used by non-experts. As the engine compiles the workflows down to Java byte code, the tight execution control and its associated features are not available.

In summary, most of these approaches come either from a rather theoretical point of view and do not address the needs of a modern IT-infrastructure and are not suitable for non-IT-experts or they do not cover the semantical layers that are necessary for using ontologies and service matching. This has been one strong motivation for creating an agent framework that tries to merge service technologies with semantical descriptions and reasoning.

3   MAMS Project

The development of new services for telecommunication applications and other IT systems is one of the most important vehicles of innovation for telecommunication and other service providers. In order to enable SMEs to utilize these new services one must find ways to accelerate the service development and deployment process. In addition we must find means to deliver those services promptly. Thus, new concepts and tools that match the requirements of SMEs are essential. Services need to be composed by non-technical persons without the knowledge of sophisticated service description and composition languages and without the need of knowing any challenging programming concepts. The MAMS project employs graphical service composition and agent-based approaches as answers to these problems.

First, a visual service creation environment for non-IT-expert users has been developed (by project partner T-Systems, DTAG) that allows the easy composition and deployment of new services [8]. Thus, the non-experts become developers and can apply their creativity and domain knowledge to literally construct new services and applications. Support of non-IT-experts has been one important driving force of the MAMS project. As a consequence, service composition requires only knowledge of input and output types of the individual services. For easy understanding, these types have been defined in a user-friendly way and include high level abstractions such as email or sip-addresses, user group names but also numbers and booleans as basic types. The user can choose from

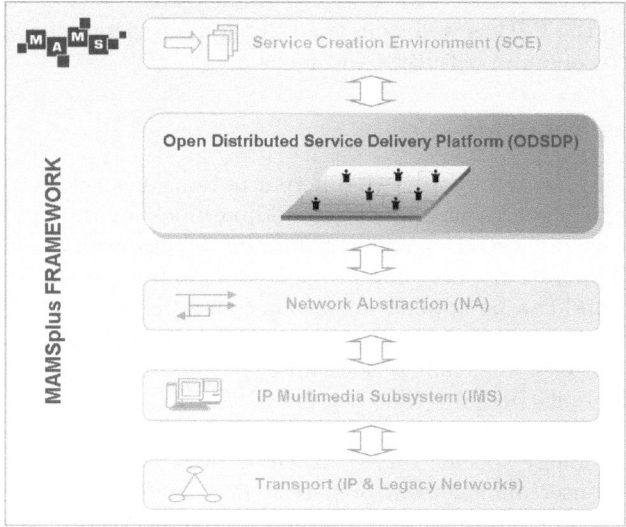

Fig. 1. MAMS Framework

a list of services and by drag 'n drop places them onto the workbench. Via a graphical representation these services are then connected to create a new service composition. In a next step, the composition will be deployed onto the ODSDP.

Second, the development tools and the service execution environment are integrated into a unified framework - the MAMS Service Framework. This allows a smooth deployment of the developed services and applications and enables a further acceleration of the development process. The service execution- as well as the development environment offer the integration of most modern infrastructures, so that services and applications can rely on up-to-date technologies and infrastructures. For example, the IP Multimedia subsystem (IMS) allows the integration of multi media connections, e.g. for video telephony or conference calls. Figure 1 shows how the MAMS framework offers a complete vertical integration of all layers, from the service creation over execution and network layer down to the hardware.

In MAMS three different roles are distinguished:

- First, the *platform provider*, which is the host and owner of the service delivery platform (SDP). This role provides and maintains basic services and manages the integration of other 3rd party services.
- Second role is the *service provider*, which is regarded as a non-IT-expert, e.g. an entrepreneur or a Small-to-Medium-Enterprise who wants to use IT-technology but does not have the resources to host and maintain his own SDP. The non-IT-expert builds its own services by combining and configuring basic services, thus creating new higher level services with the help of MAMS.

– Finally, the *service user*, which uses the higher level services. These services might be in the B2B as well as in the B2C domain, depending on the service providers (economic) activities and focus.

3.1   Services in MAMS

In the MAMS framework we use two different notions of services. The first is the *basic service* which represents the basic building block for the modelling of user generated service orchestrations. These building blocks are typically equivalent to web service operations, i.e. they provide a single functionality which is easy to understand and can be applied in various contexts.

An important part of the MAMS project was to identify and specify such simple basic services in a way, that is both understandable for a non-IT-expert and reusable in different contexts. (See Section 5) The list of basic service includes domain specific services, such as the handling of health related data (blood pressure, etc.), technology specific services such as IMS-functionalities, and generic services such as user management or persistence. In order to ensure that these different services work well together, we had to give detailed instructions to the developers of these basic services, as to how the basic services must behave and how they have to be described.

It is up to the service provider (i.e. the non-IT-expert) to combine these basic services into a *service composition* and thereby create a new application. This is the other type of service we talk about. Once this service composition is complete, it can be deployed on the ODSDP and is then available to the service users. The service creation process starts with graphical Service Description Language (SDL) which is a data flow oriented modelling language that can be used by the non-IT-expert for service compositions. From this, we generate a BPEL based intermediate model, that is used for storage of service compositions and ensures compatibility with other technologies. Finally, we translate this to a Declarative Formal Language (DFL) which is interpreted by the agents and includes ontology based declarative service descriptions. These allow the application of reasoning and service matching to the composition.

The MAMS Language Stack. The framework for modelling, deploying and executing service compositions is tripartite. It consists of:

– A graphical Service Description Language (SDL) which is a data flow oriented modelling language that can be used by the non-IT-expert for service compositions.
– A BPEL based intermediate model, that is used for storage of service compositions and ensures compatibility with other technologies by being transformable into other execution language or frameworks.
– A Declarative Formal Language (DFL) which is interpreted by the agents and includes ontology based declarative service descriptions which allow the application of reasoning and service matching to the composition. See section 4.2 and 4.1 for further details.

While the intermediate model can be executed with a normal BPEL-engine if the appropriate web services are provided, it also contains an OWL-S description for each basic service. Thus, in the DFL we can replace the WSDL-call with an abstract call to the OWL-S [2] description. The DFL script is then deployed into an agent on the ODSDP that is able to execute this script. However, as that agent now has a semantic description of the basic services that shall be executed, it is able to apply reasoning and service matching to the service composition in order to optimize the service execution. (See Section 4.1)

For example, the composition may only contain a call to an abstract 'messaging-service' that states that a message should be delivered to a certain user. The agent is then able to decide at runtime, which mechanism can be used to delivering the message, e.g. SMS or eMail, depending of the available addresses of the user or available preferences.

Furthermore, in case of a missing or failed basic service, the agent may be able to find similar services that achieve the same results, but are not explicitly referenced in the service composition. Thus, we are able to use the agents intelligence to improve the execution of the service composition.

3.2   Agent Technology in MAMS

The MAMS project utilises agent technology to realise an Open Distributed Service Delivery Platform (ODSDP) as a service integration and execution layer. This core platform connects services provided by the network abstraction (e.g. IMS based telecommunication functionalities), basic services on the platform itself for handling standard tasks (e.g. user management, basic user interface functionalities) and composed services created by the service providers in a common orchestration and execution layer. Agent Oriented Technologies where chosen to provide a range of different aspects for the development and provisioning of services. Beside using the agent metaphor as an intuitive way for modelling software components in the platform, the multi-agent system (MAS) has a range of advantages. Agents act as service providing entities encapsulating code, data and identity of a service and it's real world provider. This allows a simple and clear mapping of services from different service providers to agents on the platform. The agents communication mechanisms are abstracted from the network and based on a message bus that handles the message communication between agents. As agents allow distribution and management at runtime, the services on the ODSDP are managed with high flexibility and can be provided regardless of structural changes in the ODSDP. This allows extending or reducing hardware resources at runtime, as the agents can locate new Agent Nodes in a network and utilise those added resources as well as newly added agents and their services. Configurable Agent Beans enable plugging different services into an agent therefore allowing simple extensions or changes to an agents services and functionalities. As agents can be added, removed and detected at runtime, functionalities can be extended or reduced during operation, too.

The basic structure of the ODSDP as shown in figure 2 is composed of a host of Agent Nodes (essentially a JVM with the basic environment for running

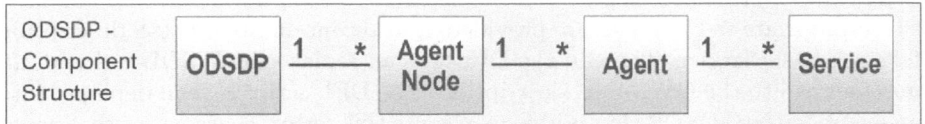

Fig. 2. ODSDP structure

agents), which each contain a number of agents. These agents include some basic functionalities and in turn a host of configurable Agent Beans. These Agent Beans are the concrete level on which functionalities like basic services are implemented by professional service developers. The strong separation of these different structural elements ensures stability and scalability in the ODSDP, as discussed later on.

While the above mentioned advantages are aspects of the system design, there are other more functional aspects of agent technology that can be usefully applied in the context of a service execution environment. One important aspect of agents is the ability for adaptive behaviour. We think that agent technology and especially multi agent systems are well suitable for creating systems that incorporate adaptive behaviour [7,15,4,6] and that adaptive agents can provide reasonable features in the context of service execution and thus enrich the MAMS service execution environment. One aspect of adaptivity in the context of service execution is the ability to seek, find and use new services in substitution of other services that might be temporary not available or not usable for some unknown reason. We have designed and successfully applied adaptive agent behaviours that deal with dynamic load balancing mechanisms as well as self healing features. Our agents have the ability to perform an intelligent service matching which enables them to substitute services at runtime and thus providing nearly seamless service usage for the end user. In addition the platform uses an adaptive load balancing method applying two basic load balancing strategies. [26]

4   Service Execution Framework

The ODSDP is an agent-based service execution environment of the MAMS framework. It consists of an arbitrary number of agent nodes which are connected by a Java Message Service(JMS) based message broker. This allows an organisational as well as a physical distribution of the services. Additional to the message broker and a comprehensive Java Management Extensions(JMX) based management interface, optional agent node components may provide infrastructure functionality, e.g. a distributed service directory connected by the message broker for announcing and searching of services in a peer-to-peer manner without a centralised database. A caching mechanism can be activated to speed up the global search. Both kinds of services (basic services and composed end-user services as described in the previous section) are provided by agents, which can be deployed on agent nodes at runtime by using the management interface as well. Advantage of this approach is that the services can be transparently distributed over different nodes, and they have their own knowledge and an own thread of control instead of being dependent on a shared engine.

For a more detailed description of the framework, please refer to [10]. In the following, we will describe the agents used in the MAMS project, as well as the relevant service provisioning and invocation mechanisms.

4.1   Agent Architecture

A MAMS agent is based on a component architecture, which always contains two standard components: a memory for the storage of an agents local knowledge and an execution cycle that controls the behaviour of the agent. Furthermore, an agent usually contains a number of optional components as shown in Figure 3. Most importantly these include communication and access to the service directory for the agent, which are in fact customary, as communication is a central feature of every multi agent system.

Other optional components give access to the service matcher, provide services or interpret service compositions. The interaction between all components inside an agent takes place via the memory according to the blackboard metaphor. In the following, we will describe the most important processes within an agent that are associated with the service provisioning and invocation.

Adaptors. An agent may have multiple adaptors, each of which can provide a list of services for the agent itself or for all agents. These adaptors are implemented with a sensor/effector model, and they are used to integrate different technologies, ranging from simple Java implementations to databases and Web Services. An adaptor registers its services with the agent during startup, or anytime thereafter. During this registration, the adaptor may either provide the agent with a complete OWL-S description of the service, which can then be used by the matcher, or it may simply provide a input/output specification, which is then wrapped in an incomplete OWL-S description which is missing preconditions, effects, etc. In either case, the resulting OWL-S description is registered at the service directory and afterwards available to all agents (unless the adaptor explicitly marked the service as local for the agent).

Service Matching. The service matching in MAMS is done by a dedicated component of the agent node. Each agent has access to this component and can submit a service template that is matched against all known services from the service directory. After the matching process, the agent receives all matching services ordered by the degree of the match from the service matcher.

For this, we constructed our own OWL-S matcher, which is able to perform a multi step matching process. With this process, we can match service descriptions with different levels of detail, ranging from simple name- and input/output matching to a semantical OWL-S matching. This is necessary, because our service directory contains different types of service descriptions, ranging from simple input/output specifications to fully fletched OWL-S descriptions. For a detailed description of how the service matcher operates, please refer to [18].

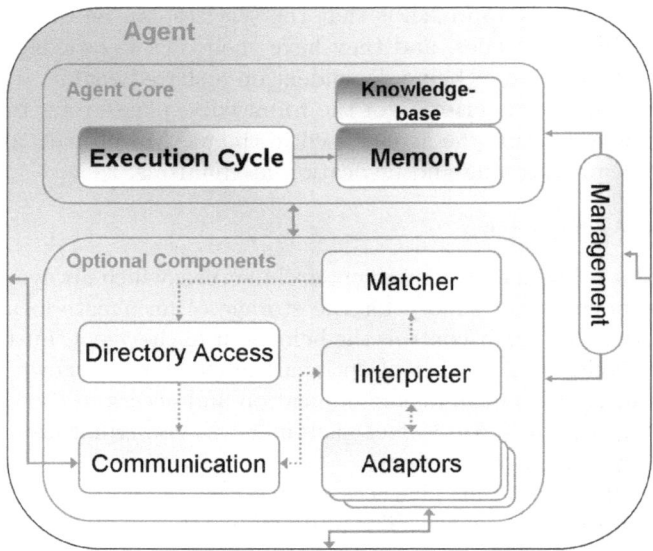

Fig. 3. A MAMS agent

Service Invocation. When an agent tries to invoke a service, that service is either provided by the agent itself or by another agent. In the first case, the agent can simply call the appropriate adaptor that registered the service in the first place and wait for the results.

The second case is more interesting. Based on the results from the service matcher, the agent will have a list of services that fulfil the requirements given by the invocation template. As these services come from the service directory, each of them has annotated the provider agent that performs the service. The agent then continues by selecting a service from the matcher list (usually the first service, as it is the best match), and sends an invocation message to the provider agent via its communication component. Please note that, though we use OWL-S descriptions for our services, we do not use SOAP for the service invocation as it is usually done. Rather, we use the JMS system that is controlled by the agent nodes. The protocol used for invocation is similar to the Request-Reply protocol, in which the user requests the service, and the provider replies with either the results or failure message.

In case of a failure, the user agent usually tries to invoke the next service on the matcher list, until either the service invocation was successful or no more matches are left. This makes our framework quite robust, as we can have multiple similar services deployed on multiple agent nodes and thus the agents can compensate for failures or malfunctions of services.

Interpreter. The interpreter can best be described as a special case of an adaptor that is able to interpret service compositions expressed in the DFL from Section 4.2. Consequently, it behaves just like an adaptor in that it registers its compositions as actions and it responds to service invocation calls just like other adaptors do.

Other than that, it is basically a simple state machine that processes a service composition step by step, executes the DFL commands, and if necessary uses the service matcher and the communication to call other services.

4.2   Service Compositions with DFL

The service composition language itself is based on the JIAC Agent Description Language (JADL) [13] and it also unites ontology based data structures and service-oriented programming. The DFL is a MAMS specific subset of JADL++, a further development of the original JADL and uses more modern techniques to reach the same goals. Details about this new language and its semantics can be found in [10].

The declarative part of the DFL is completely based on OWL [17], including OWL-S [2] for the service descriptions. This allows us to resort to existing reasoning and service matching implementations and concentrate on the integration into our agents.

The new feature of the scripting language however is the notion of *abstract service invocation*. Each invocation of a basic service within a service composition (i.e. a DFL script) is represented by the invocation of a service template. This service template is basically an OWL-S description that may be incomplete. This approach has the advantage that the same mechanism can be used for precise service invocation (if the parts of the template that are provided identify

```
<Script>        = <Decl> |
                  <Skip> |
                  <Assign> |
                  <Seq> | <Par> | <Protect> |
                  <Write> | <Read> | <Remove>
                  <IfThenElse> | <Loop>
                  <Invoke> |
                  <Send>
<Decl>          = var x: Type
<Assign>        = x := <Expr>
<Seq>           = <Script> ; <Script>
<Par>           = <Script> || <Script>
<Protect>       = protect <Script>
<Write>         = write <Expr>
<Read>          = read <Expr> x
<Remove>        = remove <Expr> x
<IfThenElse>    = if b then <Script>
                       else <Script>
<Loop>          = while b do <Script>
<Invoke>        = invoke name in: [x] out: [y]
                       pre: [A] post: [B] (nonFunc)*
<Send>          = send addr <Expr>
```

Fig. 4. DFL Abstract syntax

the service unambiguously) or for a goal driven approach. In the latter case, the template is regarded as a goal and the agent tries to find and execute a service that matches the template. By using the preconditions and effects that are given in an OWL-S description, we can improve the matching process, verify the results, and even apply planning from first principles if appropriate.

Another advantage of using an interpreted scripting language is that we can fully control the execution of a service composition. This means that we can stop the execution and resume it later, at any point of the composition. The fact that each service composition is deployed in its own agent greatly enhances the possibilities for monitoring and management of the ODSDP.

4.3   Agent Node Functionalities

In our framework, agent nodes are responsible for the management of agents as well as the control of unique resources on a machine, such as communication, databases or web servers. As a complete description of the available features of these agent nodes would be beyond the scope of this paper, we will focus on the mechanisms for the discovery and deployment of services, as these are the most interesting in this context.

Service Directory. The service directory is an mandatory component for an agent node. It provides access to the global service directory of all agent nodes to the local agents. It can be described as a proxy-solution for accessing the service directory that takes care of propagation and synchronisation of offered services autonomously.

While the general structure of the service directory integration in our architecture is quite flexible and supports multiple solutions for the implementation of the directory, we decided to use a peer-to-peer solution for the MAMS-Framework. In this approach, the agent nodes promote themselves via a multicast protocol, and afterwards exchange their local service descriptions as necessary. This approach has the advantage that the list of known service descriptions on each node is always up to date, while at the same time, we do not need any central database (such as a LDAP-directory), which may become a bottleneck.

The actual functionalities of the service directory are however nothing special. It mainly implements the operations of the white and yellow pages services described in the FIPA-DF specifications, such as registration/deregistration of agents and service and the search for both.

Service Deployment. The management interface of an agent node allows the deployment of new services during runtime. There are two possibilities to do this. One can provide an appropriate library together with the configuration of an agent that uses this library. This agent is then instantiated by the agent node and afterwards registers and provides its services via the directory and communication facilities.

The other possibility consists of a service composition that uses so called basic services in order to provide a higher level service. These compositions are expressed in the DFL we presented in Section 4.2. In order to deploy such a service composition, one must use a configuration for a standard agent that contains the corresponding interpreter. The composition itself is added as a configuration parameter to the interpreter. Afterwards, the configuration is given to the agent node, which then creates a new agent from that configuration. The interpreter of that agent then registers the service composition just like any other service during the startup cycle, and afterwards it can be used by other agents just like the basic services.

For basic services that are referenced by the service composition, the interpreter employs the service matcher whenever they are called, in order to find the best basic service that is currently available on all known agent nodes. Thus the service compositions can make use of the dynamic deployment of services on the platform.

Agent Migration. In order to further extend flexibility of agents on the distributed ODSDP runtime system agents can be migrated between different Agent Nodes, physically relocating services at runtime. The current implementation provides weak migration which transfers code and data but not the state of an agent and it's components to a different agent node. This allows to transfer services at runtime but not their active sessions or data collected during their lifetime, unless this data is kept persistent in an external repository. An implementation extending this to strong migration including code, data and state is currently in progress. Migration can be applied e.g. for administrative purposes (e.g. moving agents to other agent nodes in order to empty and later shut down a given Agent Node) or load balancing, where running agents and their services can be moved from one Agent Node with high load to other less loaded Agent Nodes, therefore optimising overall system load on the ODSDP [26,25]. Migration is realised by transferring the configuration as well as the libraries of the agent to the destination Agent Node where a specialised version of the service deployment (see above in Section 4.3) is being used.

Agent Persistence. As the deployment of new service compositions at runtime is a core element of the MAMS project. But Service Providers do not want to manually redeploy their services, after one or more Agent Nodes, that are hosting agents with composed services, have been shut down. These Agent Node configurations must not be lost and therefore the Agent Persistence implementation allows storing runtime states as configuration files of an Agent Node in the file system. These configurations can be used to restart Agent Nodes with exactly the same agents and services as their last running instances. The configuration file saved is human readable and can be edited for administrative purposes when needed, in order to manually change the configurations contained therein. The Agent persistence primary component monitors all additions or removals of agents on a given Agent Node, as well as important attribute changes. These changes are applied to an in memory representation of the configuration file which is saved to disk when the agent node is shut down. This approach allows flexible usage of saved configurations for different purposes as well as the future extension of these mechanisms to a checkpointing system for Agent Node configurations.

5   Experiences and Results

During the MAMS project, we implemented several prototypical scenarios from the eHealth and eBusiness domains in order to verify our approach. The scenarios were used to evaluate the complete processes defined in MAMS for the creation of services by non-IT-experts, including service composition, deployment and service execution. Within these scenarios we have defined several basic services and type definitions (high level input and output types of basic services) in an iterative fashion. It showed that in order to have a functionally compatible set of basic services, these types need to be defined carefully. They need to contain sufficient high level types like telephone or email addresses, which are understandable for non-IT-experts as well as generic types like text and numbers, such that input from different sources can be delegated to numerous basic services. This ensures that a rich diversity of service compositions can be created.

Furthermore, these scenarios showed us, that we needed a careful consideration of the range of abilities that are expected from a basic service. If single basic services tend to become larger and larger they can hardly be regarded as basic services any more. This restricts them from being reused. On the other hand, making a basic service too simple leaves too much composition logic to be developed by the non-technical-expert, which is conflicting with the projects goal of easy service creation. Thus basic services should have minimal functionality with some convenience logic embedded such that a non-IT-expert can easily understand their function and reuse them in various contexts. Nevertheless, there is a tradeoff between understandability and function.

When combining basic services, it is sometimes necessary or at least useful, to have basic services that can take an arbitrary number of input ports. For instance if one wants to compile a text message with input from different sources. This contradicts the concept of defined input ports but is clearly of advantage when designing new service compositions. While one can always model each service with n input ports as a sequence of a combination of the same service with 2 input ports, this makes the service combination quite bloated and makes things more complex rather than making them simpler. We think that at least two types of basic services exist, services with statically defined input/output ports and those that can have a dynamic number of input/output ports. The later are hard to model though and won't fit in predefined service description schemes.

Besides dynamic input/output ports we found that sometimes it is also useful to have generic types for input or output ports. An example will illustrate this: connecting an output port of type number with an input port of type text is not possible but sometimes wanted (i.e. when compiling a text message that contains the number of group members or a date etc.). In general type compatibility can be achieved through port type transformation services, yet again it makes service composition for non-it-experts complicated and moreover it introduces programming concepts at the level of service composition. Intuitively a non-IT-expert would expect to be able to connect any port with any other since without knowing anything about programming concepts it would make sense to him. Following this approach the basic service repository should contain services with generic input ports - *anytype ports* - which allow users to freely combine services. Of course there must be a definition of how the type transformation should be performed programmatically and we found that in most cases it is sufficient to transform input ports into the type *string*.

5.1   ODSDP

The agent framework itself proved to be advantageous when deploying new services. While the first service compositions where very small, it still was a good test for the agent framework, as we had the first external service being deployed and executed on the platform. We were able to quickly deploy both, basic services and service compositions, on a running platform. Stability and scalability of all components could be ensured due to the strict separation provided by the agent metaphor. This proved to be especially valuable when deploying and testing new services compositions, as errors in those did not affect the whole platform. Defective agents could simply be removed from the platform through the management interface. Errors did appear as a consequence of the iterative development approach. Sometimes compositions included errors such as calls to outdated or undefined basic services. Others resulted from wrong input parameter types due to faulty language transformations. However, non of the encountered errors would compromise the whole platform but only the local agent.

Over time several basic services and compositions have been deployed at the ODSDP, currently we have over 60 different basic services running while the number of service composition varies. The platform itself run stable for several months, which we regard as very good performance for a prototype platform. In return this shows that an agent based service delivery platform, as described in this paper, supports service compositions for non-IT-experts. They benefit from underlying technologies such as distributed service directory and interpreted agent programming language among others. The management features of the platform allow high-level functions that are both understandable and executable by non-IT-experts.

6   Conclusion

Within the MAMS project, we have implemented multiple scenarios from the eHealth and eBusiness domains, in order to verify our approach. These scenarios were used to evaluate the complete processes defined in MAMS for the creation of services by non-IT-experts, including service composition, deployment and service execution. Ongoing testing shows that the features of a service delivery platform like the ODSDP can be efficiently based onto a service execution environment with the help of agent-oriented software design. We have applied different technologies from the field of SOA and agents and successfully demonstrated its practicability in the context of service delivery platforms. By assigning each new service composition to a single execution agent, runtime deployment and control of service execution could be achieved. We conclude that multi agent technology is of great benefit in the design and implementation of a service execution environment as shown in this paper.

6.1   Future Work

We are currently working on three different scenarios that include much more basic services (up to 30, but services may be used more than once in a single composition). These will serve as another testing ground for the concepts and implementations of the MAMS framework. In addition, a test phase will be launched, where real users, e.g. non-IT-experts from the health care business, get the chance to evaluate the framework and to make their own experiences with the creation of services with MAMS. The results will be integrated in the next development cycle.

Currently, any ODSDP is bound to its hardware resources. When platform load gets high enough, no mechanism is able to prevent that the system runs out of resources and thus is not able to scale. We are working on mechanisms (in cooperation with project partner Alcatel-Lucent) to overcame this restriction. By using an adaptable hardware layer which is able to provide additional resources, we are able integrate new agent nodes on request. Thus, if overall platform load exceeds a certain threshold, we can simply request a new agent node from the underlying hardware layer. Due to the peer-to-peer design of the agent node infrastructure, this node will seamlessly integrate itself into the existing platform.

Most of the mechanisms described in this paper are not visible to the interested audience, besides a whole lot of log file messages. Therefore, we want to create a tool - a 3D platform monitor - that lets the user choose the level of detail. This 3D world must contain agent nodes, agents, adaptors and communication links each of them represented according to a special metaphor to visualise its key concept and to enable different levels details. In a first version the monitor envisioned shall be able to visualise platform components as shown in Figure 2, deployment/undeployment of new agents (services), inter-node and inter-agent communication, service invocation and status information of each component.

Acknowledgements. We would like to thank the Competence Centre Agent Core Technologies for their support.

This work has been sponsored by the Federal Ministry of Education and Research. (Project funding reference number 01BS0813).

References

1. Alty, J.A., Griffiths, D., Jennings, N.R., Mamdani, E.H., Struthers, A., Wiegand, M.E.: Adept - advanced decision environment for process tasks: Overview & architecture. In: Proc. BCS Expert Systems 1996 Conference, pp. 5–23 (1994)

2. Barstow, A., Hendler, J., Skall, M., Pollock, J., Martin, D., Marcatte, V., McGuinness, D.L., Yoshida, H., Roure, D.D.: OWL-S: Semantic Markup for Web Services (2004), http://www.w3.org/Submission/OWL-S/

3. BEA Systems Inc. BEA weblogic SIP server, http://www.bea.com/framework.jsp?CNT=index.htm&FP=/content/products/weblogic/

4. Bozzo, L., Mascardi, V., Ancona, D., Busetta, P.: CooWS: Adaptive BDI agents meet service-oriented programming. In: Isaias, P., Nunes, M.B. (eds.) Proceedings of the IADIS International Conference WWW/Internet 2005, vol. 2, pp. 205–209. IADIS Press (2005)

5. Caire, G., Gotta, D., Banzi, M.: WADE: A software platform to develop mission critical applications exploiting agents and workflows. In: Berger, M., Burg, B., Nishiyama, S. (eds.) Proc. of 7th Int. Conf. on Autonomous Agents and Multiagent Systems (AAMAS 2008) — Industry and Applications Track, pp. 29–36 (May 2008), www.ifaamas.org

6. Casella, G., Mascardi, V.: From AUML to WS-BPEL. Technical Report DISI-TR-06-01, Dipartimento di Informatica e Scienze dell'Informatione, Università di Genova (2006)

7. Dickinson, I., Wooldridge, M.: Agents are not (just) web services: Considering BDI agents and web services. In: Proceedings of the 2005 Workshop on Service-Oriented Computing and Agent-Based Engineering (SOCABE 2005), Utrecht, The Netherlands (July 2005)

8. Freese, B., Stein, H., Dutkowski, S., Magedanz, T.: Multi-access modular-services framework — supporting smes with an innovative service creation toolkit based on integrated sdp/ims infrastructure. In: ICIN 2007 Bordeaux (2007)

9. Greenwood, D., Calisti, M.: Engineering web service - agent integration. In: 2004 IEEE International Conference on Systems, Man and Cybernetics, vol. 2, pp. 1918–1925 (2004)

10. Hirsch, B., Konnerth, T., Heßler, A.: Merging agents and services — the JIAC agent platform. In: Bordini, R.H., Dastani, M., Dix, J., El Fallah Seghrouchni, A. (eds.) Multi-Agent Programming: Languages, Tools and Applications, pp. 159–185. Springer, Heidelberg (2009)

11. HP. HP service delivery platform, http://h71028.www7.hp.com/ERC/downloads/4AA0-7670ENW.pdf

12. IBM. IBM service provider delivery environment, http://www-03.ibm.com/industries/telecom/doc/content/solution/257531102.html

13. Konnerth, T., Hirsch, B., Albayrak, S.: JADL — an agent description language for smart agents. In: Baldoni, M., Endriss, U. (eds.) DALT 2006. LNCS, vol. 4327, pp. 141–155. Springer, Heidelberg (2006)

14. LOMS. Local mobile services, http://www.loms-itea.org

15. Luck, M., McBurney, P., Shehory, O., Willmott, S.: Agent Technology: Computing as Interaction (A Roadmap for Agent Based Computing). AgentLink (2005)

16. MAMS. MAMS multi access - modular services, http://www.mams-platform.de/

17. Martin, D., Hodgson, R., Horrocks, I., Yendluri, P.: OWL 1.1 web ontology language (2006), http://www.w3.org/Submission/2006/10/

18. Masuch, N.: Development of a standard-based service matcher component within an agent-oriented framework. Diploma thesis, Technische Universität Berlin (2008)

19. Microsoft. Microsoft connected services framework, http://www.microsoft.com/serviceproviders/default.mspx

20. O'Brien, P.D., Wiegand, W.: Agent based process management: Applying intelligent agents to workflow. The Knowledge Engineering Review 13(2), 1–14 (1998)
21. OPUCE. Open platform for user-centric service creation and execution, http://www.opuce.tid.es
22. Preuveneers, D., Pauty, J., Landuyt, D.V., Berbers, Y., Joosen, W.: Comparative evaluation of converged service-oriented architectures. In: AINA Workshops, vol. (1), pp. 989–994. IEEE Computer Society, Los Alamitos (2007)
23. Singh, M.P., Huhns, M.N.: Multiagent systems for workflow. International Journal of Intelligent Systems in Accounting, Finance and Management 8, 105–117 (1999)
24. SPICE. Service platform for innovative communication environment, http://www.ist-spice.org
25. Stender, J.: Grid-inspired load balancing in service-based multi-agent systems. Master's thesis, Technische Universität Berlin (August 2005)
26. Thiele, A., Konnerth, T., Kaiser, S., Keiser, J., Hirsch, B.: Applying JIAC V to real world problems — the MAMS case. In: Proceedings of the German conference on Multi-Agent System Technologies, Springer, Heidelberg (to appear, 2009)
27. Walton, C.: Uniting agents and web services. In: Agentlink News, vol. 18, pp. 26–28. AgentLink (2005)
28. Zinnikus, I., Hahn, C., Fischer, K.: Model-driven, agent-based approach for the integration of services into a collaborative business process. In: Pagham, Parkes, Müller, Parsons (eds.) Proc. of 7th Int. Conf. on Autonomous Agents and Multi-agent Systems (AAMAS 2008), Estoril, Portugal, May 12-16, 2008, pp. 241–248 (2008)

Author Index